TRANSACTIONS

of the

AMERICAN PHILOSOPHICAL SOCIETY

Volume 90, Pt. 2

TRANSACTIONS
of the
AMERICAN PHILOSOPHICAL SOCIETY
Held at Philadelphia
For Promoting Useful Knowledge
Volume 90, Pt. 2

Astronomy in the Iberian Peninsula:

ABRAHAM ZACUT AND THE TRANSITION FROM MANUSCRIPT TO PRINT

JOSÉ CHABÁS and
BERNARD R. GOLDSTEIN

American Philosophical Society
Independence Square • Philadelphia
2000

Library of Congress Cataloging-in-Publication Data

Chabás, José, 1948–
Astronomy in the Iberian Peninsula : Abraham Zacut and the transition from manuscript to print / José Chabás and Bernard R. Goldstein.
p. cm.—(Transactions of the American Philosophical Society, ISSN 0065-9746 ; v. 90, pt. 2)
Includes bibliographical references and indexes.
ISBN 0-87169-902-8 (paper)
1. Zacuto, Abraham ben Samuel, b. ca. 1450. Almanach perpetuum celestium motuum. 2. Astronomy—Portugal—History—To 1500. I. Goldstein, Bernard R. II. Title. III. Series.
QB36.Z3 C48 2000
520'.94609'024—dc21 00-023028

ISBN:0-87169-902-8
US ISSN: 0065–9746

To Clara

Contents

List of Figures ix
Preface xi

Introduction 1

1. Abraham Zacut: Supplementary notes for a biography 6

2. Setting the Scene 16
   - 2.1 The almanac tradition 16
   - 2.2 The Alfonsine Tables and Salamanca 18
   - 2.3 The *Tabule Verificate* for Salamanca 23
   - 2.4 The "Tables in Castilian" 37
   - 2.5 Other material related to Salamanca 47
   - 2.6 Predecessors explicitly acknowledged by Zacut 49

3. The *Ḥibbur* 53
   - 3.1 The tables 53
   - 3.2 Other tables related to the *Ḥibbur* 81

4. The *Almanach Perpetuum* 90
   - 4.1 The dedication 90
   - 4.2 The canons 95
   - 4.3 The tables 98
   - 4.4 Dating the publication of the *Almanach Perpetuum* 156
   - 4.5 Joseph Vizinus 156

5. The Influence of Zacut's Astronomical Works 161
   - 5.1 Disciples 161
   - 5.2 Later editions of the *Almanach Perpetuum* 161
   - 5.3 Zacut's impact on the Jewish community 163
   - 5.4 Zacut's impact on Christian scholars 164
   - 5.5 The *Almanach Perpetuum* in the Muslim world 170

Appendix 1. Zacut's *Judgments of the astrologer* 173
Bibliography 175
Notation 183
Indices: 185
   - Manuscripts Cited 187
   - Parameters 191
   - Names and Subjects 193

List of Figures

Fig. 1. Colophon, *Almanach Perpetuum*, Leiria, 1496, f. 172r (Madrid, Biblioteca Nacional, shelfmark I–1077). 10

Fig. 2. Document with Zacut's signature (Lisbon, Arquivo Nacional da Torre do Tombo, *Corpo Chronologico*, parte 1ª, maço 2, doc. 18). 12

Fig. 3. Table TV 4: Table for the deficit with respect to an integer number of days (Madrid, Biblioteca Nacional, MS 3385, f. 104v). 27

Fig. 4. Table TC 1: Table for the mean motion in elongation (Madrid, Biblioteca Nacional, MS 3385, f. 139r). 38

Fig. 5. Table HG 1: Table for the true position of the Sun: Epoch 1473, first page only (Madrid, Biblioteca Nacional, MS 3385, f. 1r). 57

Fig. 6a. Dedication, *Almanach Perpetuum*, Leiria, 1496, f. 2r (Madrid, Biblioteca Nacional, shelfmark I–1077). 91

Fig. 6b. Dedication, *Almanach Perpetuum*, Leiria, 1496, f. 2v (Madrid, Biblioteca Nacional, shelfmark I–1077). 92

Fig. 7. Table AP 2: Table for the daily solar positions, first page only, *Almanach Perpetuum*, Leiria, 1496, f. 21r (Madrid, Biblioteca Nacional, shelfmark I–1077). 105

Fig. 8. Table AP 7: Table for the daily lunar positions, first page only, *Almanach Perpetuum*, Leiria, 1496, f. 30r (Madrid, Biblioteca Nacional, shelfmark I–1077). 114

Fig. 9. Table AP 12: Solar elongation from the lunar node, *Almanach Perpetuum*, Leiria, 1496, f. 63r (Madrid, Biblioteca Nacional, shelfmark I–1077). 119

Fig. 10. Table AP 23: Latitude of Saturn, first page only, *Almanach Perpetuum*, Leiria, 1496, f. 77v (Madrid, Biblioteca Nacional, I–1077). 139

Fig. 11. Table AP 42: Unequal motion of Mercury, first page only, *Almanach Perpetuum*, Leiria, 1496, f. 149r (Madrid, Biblioteca Nacional, I–1077). 146

Fig. 12a. Table AP 45: Fixed stars *Almanach Perpetuum*, Leiria, 1496, f. 164v (Madrid, Biblioteca Nacional, I–1077). 148

Fig. 12b. Table AP 45: Fixed stars *Almanach Perpetuum*, Leiria, 1496, f. 165r (Madrid, Biblioteca Nacional, I–1077). 149

Fig. 13. Title page: *Almanach perpetuum exactissime nuper emendatum omnium celi motuum cum additionibus in eo factis tenens complementum.* (Venice: Petrus Liechtenstein, 1502). 151

Preface

Our interest in the astronomical traditions of the Iberian Peninsula began with our respective doctoral dissertations, and has continued ever since. Our collaboration goes back to the research leading to the publication in 1992 of a joint paper on a remarkable 'user-friendly' astronomical table by Nicholaus de Heybech (ca. 1400). Of particular interest to us was that this author did not invent any new theories and did not depend on any new observations but, working within the Alfonsine tradition, he eliminated the need for complicated computations while still achieving the desired accuracy. This showed us that significant contributions to astronomy need not involve alterations in fundamental theories or new observations, matters that have been the focus of attention for most historians of astronomy. Here we apply this insight into our analysis of Zacut's astronomy.

Since Latin astronomy in the late Middle Ages is mainly about the Alfonsine Tables (named after Alfonso X of Castile, d. 1284), it came as a surprise that we could find no trace of them in astronomical tables produced in Spain until the 1460s. Moreover, the (re)introduction of these tables depended on the arrival in Salamanca of a Polish astronomer who brought with him Alfonsine material from Cracow. Zacut benefited from contact with Christian astronomers in Salamanca who in turn depended on a vast corpus produced by astronomers all over northern Europe (from Oxford to Cracow, and many points in between). He also took advantage of the little known, but impressive, Jewish tradition in astronomy that developed mainly in southern France and Spain in the late Middle Ages. We were prepared to recognize these various elements in Zacut's tables because of our previous experience with both Latin and Hebrew sources as well as the underlying Greek and Arabic traditions.

In addition to Latin and Hebrew, our work has also led us to consult medieval texts in Castilian, Catalan, Portuguese, as well as Arabic, and we shared the burden of reading them according to our skills. But, in all respects, this has been a collaborative work, and all matters have been thoroughly discussed, often by e-mail, and sometimes face-to-face. We have not allowed the distance between Europe and the USA, our different mother-tongues, or our dependence on different computers (mac vs. windows) to inhibit our communications.

We thank the librarians at many institutions for providing copies of manuscripts and early printed books, some of which we have also consulted directly. The Biblioteca Nacional de Madrid and the Arquivo Nacional da Torre do Tombo (Lisbon) kindly granted us permission to include photographs of documents in their possession. We are also grateful to the

American Philosophical Society for its willingness to publish this analysis of Zacut's work and, more generally, for its encouragement of technical studies in the history of astronomy. Finally, we are indebted to our editor, Carole LeFaivre-Rochester, for guiding our manuscript through the various stages of publication.

We are also grateful for the cooperation and assistance of many colleagues: Peter Barker for supplying us with copies of the editions of Regiomontanus's *Tabulae Directionum*; Alan Cooper, Edna Engel, Miriam Fraenkel, Daniel Frank, Ruth Glasner, Giora Hon, Richard Judd, Y. Tzvi Langermann, and Benjamin Richler for advice on matters pertaining to texts in Hebrew; Jerzy Dobrzycki, José Luis Mancha, John D. North, and Beatriz Porres for advice on matters concerning texts in Latin; David A. King and Julio Samsó for help with the Arabic versions of the *Almanach Perpetuum*; David Herald and Peter Huber for providing us with astronomical data from modern computer programs; and Luís García Ballester for offering detailed comments on an earlier version of this work. We thank our respective universities (Universitat Pompeu Fabra, and the University of Pittsburgh) for their support of historical research and for providing favorable conditions for our collaboration.

José Chabás and Bernard R. Goldstein
Pittsburgh, November 1999

Introduction

Abraham Zacut (1452–1515) of Salamanca was an outstanding intellectual figure in the Spanish Jewish community on the eve of the expulsion in 1492. His scientific work began in the 1470s, and continued in exile, in Portugal, North Africa, and ultimately in Jerusalem. In this monograph we shall focus on some of his important contributions to astronomy, namely, those that appear in the book published in Leiria, Portugal, in 1496, generally known as the *Almanach Perpetuum*, and we will be careful to distinguish this publication from *ha-Ḥibbur ha-gadol* (*The Great Composition*) that Zacut composed in Hebrew in 1478. Indeed, one of our findings in the course of this research is that these are distinct works, and that the *Almanach Perpetuum* should no longer be considered a translation of Zacut's *Ḥibbur*. As we shall also argue, Zacut was well aware of developments in astronomy among his Christian contemporaries in Salamanca and depended on them to a significant degree, even though he does not reveal it openly. In contrast to Zacut's astronomy that was published in his lifetime (although he may have had nothing to do with the publication itself), the works on mathematical astronomy by his Christian contemporaries in Salamanca have only been preserved in manuscripts.[1] Yet, even when we take into consideration a wider range of astronomical materials from the late fifteenth century than what is available in print, it is abundantly clear that Zacut was pre-eminent among astronomers in Spain at the time.

The most complete biography of Abraham Zacut is still that written by Francisco Cantera Burgos (1935, pp. 7–39). It has not been our intention to produce another biography, but only to offer a critical reading of some of the materials previously published concerning Zacut. Cantera's examination of Zacut's life and works, published separately in 1931 and 1935, is a model of careful scholarship, and we owe him a great debt of gratitude for his efforts. Cantera published many documents, including an edition of Juan de Salaya's Castilian version of the canons to the *Ḥibbur*, completed in 1481 with the aid of the author, as we learn from the colophon. Cantera also compared Hebrew manuscripts of the canons to the *Ḥibbur* with the Castilian version of these canons by Salaya. Because of Cantera's successful treatment of Zacut's text (despite a few divergences among the manuscripts that he failed to notice, some of which are not found in the manuscripts he used), we focus on the tables and their mathematical structure, rather than on the canons.

[1] For a list of incunabula on astronomy and related disciplines written by Spaniards, see Chabás and Roca 1998, p. 134, n. 6.

For well over a century, Zacut has been associated with the period of the great discoveries, and it is usually said that he played a significant role in educating Portuguese navigators. We are skeptical of these claims that depend on sources of dubious reliability, and we insist that Zacut produced a book on astronomy, not on navigation. It should be noted that Zacut never lived near the sea before his arrival in Portugal in 1492 or 1493 and that, in his extant works, he never discusses astronomical instruments or problems of astronomical navigation. Indeed, he recorded few astronomical observations, and mentions no instruments in his descriptions of them.

The *Almanach Perpetuum* consists of a set of canons that take up relatively few pages, followed by a great number of astronomical tables for various purposes. The canons are different from those in the *Ḥibbur*, but the tables were largely taken from it. Most of the tables are in the form of an almanac, that is, a set of positions for a given planet (including the Sun and the Moon), arranged at intervals of a day or a few days over the period of the planet's motion (ranging up to 125 years, in the case of Mercury). Using modern calculators, we have verified that Zacut accurately computed the entries in these tables from the Alfonsine Tables; to do this work by hand required an enormous effort, and to do it correctly required a high level of skill and careful attention to detail. For example, the lunar table required the computation of 11,325 consecutive daily positions (about 31 years), based on the full lunar model as presented in the Alfonsine Tables. The advantage of an almanac is that it is "user-friendly," requiring only linear interpolation between adjacent entries. In addition to analyzing Zacut's tables, we trace the antecedent tradition of almanacs in the Iberian Peninsula in order to convey the fact that this tradition culminates with Zacut. The user-friendliness of Zacut's tables is also evident in the case of planetary latitudes, for he presents double argument tables, once again only requiring linear interpolation. All this is done without challenging the theoretical basis of the Alfonsine Tables, or discussing the underlying models.

Surprisingly, although the Alfonsine Tables were produced in Castile in the thirteenth century at the court of Alfonso X, the earliest evidence for their use in Spain comes from *ca.* 1460 in Salamanca, shortly before Zacut began his astronomical activity, with the arrival of Nicholaus Polonius, the first incumbent of the chair of astronomy/astrology at the university there. From this time on there was a lively tradition in astronomy at Salamanca and, for various reasons, we believe that Zacut was aware of it: the most important evidence will be presented in chapter 2. It is noteworthy that Zacut was ready to use the latest material, even if he does not acknowledge the relevant sources.

Zacut was also heir to a long and distinguished astronomical tradition in Hebrew, and he acknowledges the works of Levi ben Gerson, Immanuel ben Jacob Bonfils, Jacob ben David Bonjorn, and Judah ben Asher II (fourteenth century); and Judah ben Verga (fifteenth century), among others. On the other hand, his information on earlier astronomical works in Arabic probably derived from references to them in either Hebrew or Latin treatises.

The edition of the *Almanach Perpetuum* of 1496, on which Zacut's fame rests, has many peculiarities, beginning with the fact that some copies have the canons in Latin, and others have them in Castilian (even though the book was printed in Portugal). It was clearly produced in a great hurry, with some tables in the wrong order, others with defective headings, and confusion in the canons (e.g., there are two chapters labeled "chapter 7" in the Castilian version of the canons). Most bizarre is the absence of star names in the star list, while preserving their positions and magnitudes. Despite this carelessness, the tables are rather well printed with relatively few mistakes in the entries. Responsibility for this edition presumably rests primarily with the printer, d'Ortas, whose other publications were exclusively Hebrew texts. Associated with d'Ortas is Joseph Vizinus, mentioned in the colophon to the Castilian version as having translated the text from Hebrew into Latin, and then from Latin into Castilian ("noestro vulgar romançe"). It is not easy to determine the extent of Vizinus's involvement in the publication, although it is clear that the canons in Latin (or in Castilian) were not translated from Hebrew. Moreover, there are reasons to doubt that Zacut was even consulted in the preparation of this edition, and he never mentions it in any later work, as far as we know. Nevertheless, in the modern secondary literature, Vizinus is given a major role in the history of astronomy and navigation, largely based on his supposed skill in astronomy demonstrated in this edition of Zacut's tables.

Printers around 1500 often operated in great haste to recover the cost of paper, labor, and typeface, that was required before any profit could be realized. Further, their relationship to authors was not fixed by law or custom, and living authors were not generally consulted about the publication of their works. So, it would seem, that d'Ortas practiced his craft in ways that were not unusual for his time. The introduction of printing made texts accessible to a wider public but, at the same time, it fixed them in a way that strongly contrasted to their diversity in the manuscript tradition. In fact, it is rare to find two manuscripts of the same astronomical text that agree exactly, even apart from minor variants. For example, it often happens that entire tables were added or omitted. Moreover, given the haste in preparing editions, the quality of the printed text did not always meet the standard of good manuscripts that were copied more leisurely. Nevertheless, it is fair to say that printing was instrumental in advancing the pace of scholarly communication, and was conducive to opening new avenues of scientific research.[2]

[2] Jardine 1996, espec. pp. 135–180; Lowry 1979, espec. pp. 30, 224 ff; Stillwell 1970, p. xii.

Judging from the standard histories of astronomy, the effect of moving from manuscript to print has been neglected. To be sure, the transformation was never complete, for manuscripts continued to be copied, particularly in Hebrew and Arabic, until well into the nineteenth century. The earliest stage in the systematic publication of mathematical and astronomical texts can be identified with the appearance in 1474 of Regiomontanus's *Tradelist* in which he indicated his intention to print a great number of scientific treatises.[3] Since Regiomontanus died two years later in 1476, he was unable to realize his ambitions, but others continued to fulfill them, and this included a series of editions of Regiomontanus's own works.

Regiomontanus is seen as an innovator, or at least a renovator, of astronomy, but little attention has been given to his role as transmitter of the antecedent tradition, and his desire to publish scientific texts. Zacut, on the other hand, had more modest ambitions, and we believe it appropriate to consider him in the context of his predecessors and contemporaries, particularly in Spain (based on primary sources), without regard to episodes that took place elsewhere in Europe later in the sixteenth century, such as the theoretical innovations of Copernicus or the observational achievements of Tycho Brahe, both of whom were indebted to Regiomontanus in one way or another. The printer, d'Ortas, had a different view, for he silently paid tribute to Regiomontanus by including in his publication of the Latin version of 1496 a dedication to an unnamed bishop of Salamanca that was barely changed from a dedication by Regiomontanus to a bishop in Hungary (printed in 1490), and some tables taken from Regiomontanus's *Kalendarium* (Venice, 1483), without acknowledging his sources. Again, we see no involvement of Zacut in what for d'Ortas was surely a business decision (see chapter 4, below).

We have also attempted to assess Zacut's influence on subsequent astronomers, but that is a task that goes well beyond the limits of this book. There was an immediate impact in Salamanca where we find texts in Latin and Castilian that are based on the *Ḥibbur* (independently of the *Almanach Perpetuum*). Christian scholars tended to display the same reticence in mentioning Zacut that Zacut had in mentioning Christians; hence it is impossible to determine the character of face-to-face contacts and the circumstances under which they took place. There were also several editions of the *Almanach Perpetuum* in Latin in the sixteenth century, attesting to its popularity, and there were at least two translations into Arabic. Zacut's influence on Jewish scholars was most notable in the Eastern Islamic world, based in large measure on the work he did in Jerusalem, shortly before his death.

[3] Rose 1975, pp. 104–110; Zinner 1990, pp. 110–119.

Finally, we wish to emphasize that the heart and soul of this book is the systematic analysis of the tables themselves, and this has allowed us to relate Zacut's work to those of his predecessors, both Jews and Christians. For us, numbers count.

I. Abraham Zacut: Supplementary Notes for a Biography

Abraham bar Samuel bar Abraham Zacut was born in Salamanca (Spain) in 1452, for he reports in a text written in 1513 that he was then 61 years old (London, MS Sassoon 799, f. 1a:24–30, and Paris, Alliance Israélite, MS VIII.E.60, f. 1a:24–31; see Goldstein 1981, p. 240; Shochat 1948–1949; Sassoon 1932, pp. 510–511). His family had previously lived in France, and in his *Sefer Yuḥasin* (*Book of Genealogies*),[1] Zacut indicates that his great–grandfather (*avi ziqni*) came from France (p. 223a). The date of Zacut's birth is not explicitly given in any of his works while he was in Spain. But in a lengthy astronomical treatise in Hebrew with both canons and tables written by Zacut and entitled *ha-Ḥibbur ha-gadol* (*The Great Composition*), we are repeatedly given some examples for a specific date, likely to be the date of his birth. The relevant passage in chapter 2 reads: "the 'native' was born in [1]452, 12 August, 3 hours after noon" (MS B, f. 11a–b; cf. MS S, f. 7r: *uno nasçio el año de 1452 a doze dias de agosto a tres horas despues del medio dia*). Cantera (1931, p. 68) wondered whether this was in fact Zacut's date of birth. This can now be confirmed because in Zacut's canons to his tables for Jerusalem of 1513, he indicates that he was 61 years old at that time and this implies that he was born in 1452 (Goldstein 1981, p. 240).

Very little is known about his studies or his teachers. The only direct evidence is given in the *Book of Genealogies*, where his father, Samuel, and R. Isaac Aboab are mentioned as his teachers (pp. 123a and 226a). It is not possible to indicate the precise relationship between Zacut and R. Isaac Aboab, but Zacut explains (p. 226a) that R. Aboab died in Portugal in 1493, 7 months after the expulsion of the Jews from Spain, and adds: "I was there and gave the eulogy for him, based on the passage 'I send an angel [before you] . . .' [Exod. 23:20]."

Although he is mainly known for his astronomical activity, Abraham Zacut wrote on other subjects such as lexicography (*Hosafot lasefer ha-'arukh*) and history (*Sefer Yuḥasin*). His first book on astronomy is

[1] The date of Abraham Zacut's *Book of Genealogies* is disputed. According to Alfred Freimann (1924, p. x), Zunz gave 5262 A.M. [= 1502], but most scholars, including Steinschneider, give 5264 A.M. [= 1504]. Freimann believed that the composition of the work was begun by Zacut long before its completion. But there is no doubt that it was written (or completed) in Tunis, as Zacut indicates (Zacut, *Sefer Yuḥasin*, p. 215a). For an explanation of manuscript sigla, see p. 53.

ha-Ḥibbur ha-gadol, consisting of canons and tables, and written in Hebrew for a Jewish audience. The epoch of the tables is the year 1473, but Zacut finished his work around 1478. Three years later the treatise was translated into Castilian, with the help of Zacut himself, by Juan de Salaya, who held the chair of astronomy at the University of Salamanca (1464–1469). The unique manuscript containing the translation, which is now at the library of the University of Salamanca (MS 2–163), was transcribed by Cantera (1931, pp. 151–236).

Some scholars have maintained that Abraham Zacut was a student or professor at the University of Salamanca, and that he even taught at universities in other Spanish cities (Zaragoza, Cartagena, etc.), but none of these claims is credible. The evidence for his relationship with the University of Salamanca comes entirely from a passage in a dedication in Latin ascribed to Zacut and preceding the *Almanach Perpetuum*, which contains a summary of the tables of the *Ḥibbur. The Almanach Perpetuum* was first published in 1496 at Leiria (Portugal) with the title *Tabule tabularum celestium motuum astronomi zacuti*. As will be shown later, this dedication has little to do with Zacut, for it was copied almost verbatim from an entirely different book printed for the first time in 1490, written by Regiomontanus, and dedicated to a different person, in an entirely different context. Nevertheless, we shall argue that Zacut was well informed concerning the study of astronomy at the University of Salamanca, judging from the close similarities of some of his astronomical tables with those preserved in Latin manuscripts produced in Salamanca at the time. His access to these tables (either in these manuscripts or others like them) is reasonably certain, but the way in which he had access to them remains unknown.

Of course, this does not exclude Zacut from having had disciples; indeed, this seems to be the case with Augustinus Ricius, who says he was Zacut's student at Carthage (which presumably refers to Tunis).[2]

According to Cantera, the dedication in *the Almanach Perpetuum* was addressed to Gonzalo de Vivero, Bishop of Salamanca, presumed to be the patron of Zacut for several years. However, it can be shown that this dedication had nothing to do with the bishop of Salamanca, or with Zacut (see below, chapter 2). The only available source for the relationship between Zacut and Gonzalo de Vivero, who died at the beginning of 1480, is the latter's testament, also transcribed by Cantera (1931, pp. 76, 391–396).

---

[2] A. Ricius, *De motu octauae sphaerae* (Trento 1513). A second edition appeared in Paris (1521), where we read: "Habraham Zacuth, astronomiae nostra tempestate peritissimus in sua magna editione affirmat, nobisque eum legentem in Africa apud Carthaginem audientibus, . . ." (f. 6v), and "Abraham Zacuth, quem praeceptorem in astronomia habuimus . . ." (f. 29r). The first sentence may be translated as follows: "Abraham Zacut, most learned in astronomy in our time, in his great composition [latin: editione, i.e. *Ḥibbur*] asserts, and we have listened to him lecturing in Tunis [*lit.* Carthage], Africa, . . ."

The bishop ordered that "the Jew Abraham, astrologer" be given a modest amount of money and wheat, and that there be bound in a single volume, to be put in his library, "the notebooks which are in Romance (language) and written by this Jew, . . . in order to understand better the tables by this Jew." Note that Gonzalo de Vivero refers to "Abraham," not to "Zacut," and that no indication of the actual relationship between these two men is given. In sum, we take this passage, which is the only evidence for their relationship, as insufficient to consider the bishop as a "generous protector of Zacut" or Zacut as a "close friend" of the bishop (Cantera 1931, p. 77).

Juan de Salaya, the professor of astronomy at the University of Salamanca, is also mentioned in the bishop's testament (Cantera 1931, p. 394); he gets three books: one on geomancy, one by Abū Maʿshar, and yet another one not specified. One year later, in 1481, he translated, with the help of Zacut, the canons of the *Ḥibbur* into Castilian.

At the request of Juan de Zúñiga y Pimentel (d. 1504), his patron and the last master of the Order of Alcántara, Zacut wrote a *Tratado breve en las ynfluencias del cielo* (Carvalho 1927; and Carvalho 1947, pp. 101 ff) in Gata, a locality in Cáceres, in the region of Extremadura, close to the Portuguese border. In the dedication we read:

> And for this reason, the most magnificent and of great lineage, my illustrious lord, the master of Alcántara, Don Juan de Çúñiga, lover (*amador*) of all the sciences and well versed in them, for whose fame all scholars and learned people leave their homeland and birthplace in order to seek true calm and total perfection, for whose sake those who master the sciences exert themselves, and [from whom] they receive nourishment and remuneration, and all the scholars can certainly say [about him] what the Queen of Sheba said about King Solomon: "You magnify your fame, blessed [are] your servants, those who hear your words" [cf. 1 Kgs 10:8]. He saw fit to order me, Rabbi Abraham Zacut of Salamanca, his servant [and] astrologer, to compose a short treatise on the influences of the heavens in order that the physicians of his lordship would benefit . . . (Carvalho 1927, p. 17).

The *Tratado breve en las ynfluencias del cielo* was written in 1486, and it is followed by a short text entitled *De los eclipses del sol y la luna*.[3] The date of the first treatise can be determined from a passage where Zacut makes reference to the remarkable solar eclipse of 1485: "last year, that is [14]85, even though the solar eclipse signified dryness . . ." (Carvalho 1927, p. 40). In the introductory paragraph of *De los eclipses del sol y la luna*, Zacut refers again to Juan de Zúñiga as "my lord," and to himself as "Rabbi Abraham astrologer of Salamanca." Both in the *Tratado*

[3] Carvalho transcribed both texts from the unique extant copy, Seville, Biblioteca Colombina, MS 5-2-21.

breve and the *De los eclipses*, Zacut mentions a previous work that is not extant, *Juyzio del eclipse* (Carvalho 1927, 40:9,14), concerning the solar eclipse of March 16, 1485.

When the Jews were expelled from Spain in 1492, Zacut moved to Portugal. Some evidence suggests that he served King João II and, after João's death in 1495, King Manuel I. Zacut remained in Portugal until 1496 or 1497, when the Jews of Portugal were forcibly converted.[4] The evidence for Abraham Zacut's activities in Portugal is meager. There are, however, various documents where his name is associated with the Portuguese monarchs. One of them is a letter dated June 9, 1493, where King João II orders the payment to "Raby Abraão, estrolico" an amount of "diez espadys douro" with no indication of the reason for it.[5] And in the colophon of the *Almanach Perpetuum* (1496) Zacut is referred to as astronomer "serenisimi Regis emanuel Rex portugalie" (see Figure 1).

Gaspar Correia,[6] the chronicler who wrote a historical account of the epoch of discoveries long after the time when Abraham Zacut was in Portugal, also mentions Zacut's name a few times (see Bensaude 1912, pp. 255–260; Cantera 1935, pp. 33–37). He claimed that King Manuel consulted Zacut before sponsoring the voyage that led to the discovery of India on weather and storms occurring during navigation, and that Zacut gave instructions to pilots concerning the use of astronomical instruments at sea.[7] It would be a task for an expert in Portuguese history of the time to decide how reliable Correia is. But the theme in Correia's account about Zacut giving instructions to navigators at sea seems dubious.[8] Zacut lived in Salamanca and later in Extremadura—far from the sea. We know of no evidence suggesting that Zacut had even been near the sea

[4] An edict of expulsion was decreed in December 1496 to be effective in October 1497 but, in fact, no Jews were permitted to leave Portugal and they were forcibly converted to Christianity. See Zacut, *Book of Genealogies* (Alfred Freimann 1924, p. 227a-b); M. A. Cohen, (1965, pp. 2–3, 202–204); and Cantera 1935, p. 37.

[5] This document is now in Lisbon, Arquivo Nacional da Torre do Tombo, *Corpo Chronologico*, parte 1ª, maço 2, doc. 18. See Viterbo (1898–1900), especially I:326 and II:285, which correspond to pp. 362, 673 in the 1988 edition.

[6] The date of his birth is not known, but there is evidence that he was in India in 1512; he probably died in 1561. Correia is the author of *Lendas da Índia*, a work published in 8 volumes in 1858–1866 in Lisbon.

[7] From the "report" by Correia concerning Zacut's vague answer (*Lendas*, book I, vol. I, chap. VIII) to a vague question by King Manuel on the voyage to India (*Lendas*, book I, vol. I, chap. III) it is simply not possible to deduce that Zacut made any astronomical tables for Vasco da Gama, or even that they ever met.

[8] Correia's account concerning the teaching of the use of nautical instruments to Portuguese pilots by Zacut is considered "suspect" by Albuquerque (1991, p. 144): "Gaspar Correia can be forgiven for this error, because he wrote his text in India, and in this passage he refers to a subject with which he is not familiar, and of which he could not be well informed, because he did not have access to archives or to anyone who could inform him better."

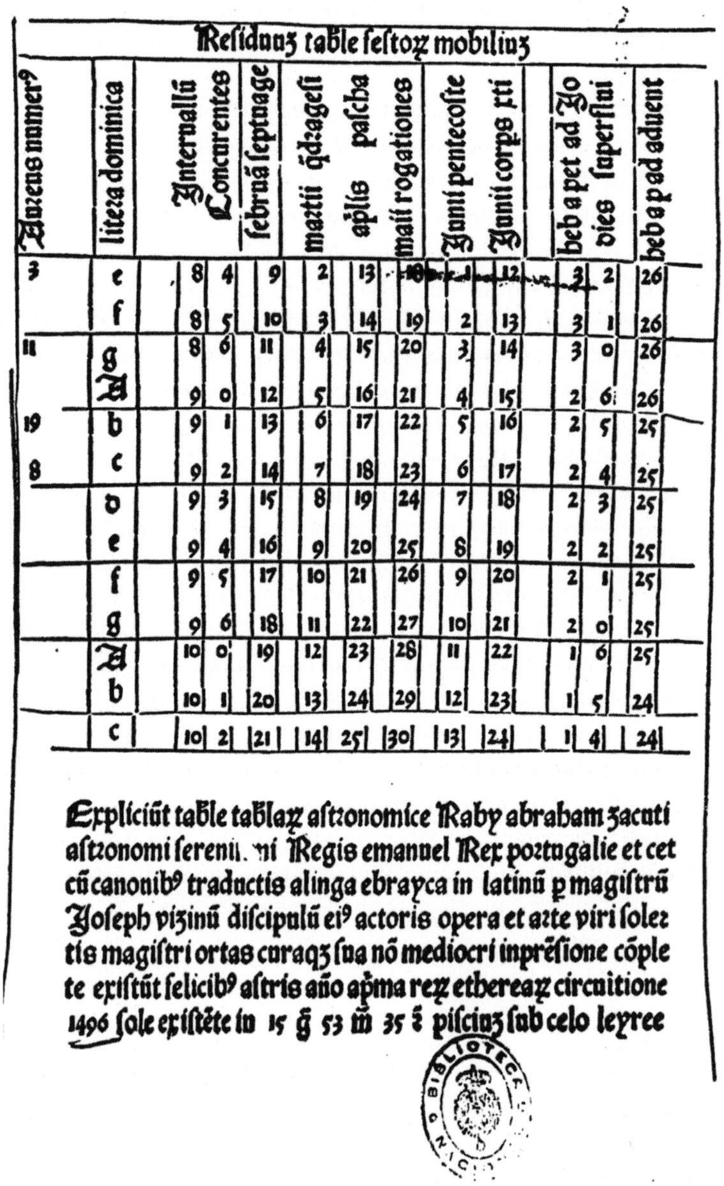

Residuum table festorum mobilium

Aureus numerus	litera dominica	Intervallum	Concurrentes	februarii septuage	martii quadragesi	aprilis pascha	maii rogationes	Iunii pentecoste	Iunii corpus xti	heb a pet ad Io	dies superflui	heb a p ad aduent
3	e	8	4	9	2	13	18	1	12	3	2	26
	f	8	5	10	3	14	19	2	13	3	1	26
11	g	8	6	11	4	15	20	3	14	3	0	26
	A	9	0	12	5	16	21	4	15	2	6	26
19	b	9	1	13	6	17	22	5	16	2	5	25
8	c	9	2	14	7	18	23	6	17	2	4	25
	d	9	3	15	8	19	24	7	18	2	3	25
	e	9	4	16	9	20	25	8	19	2	2	25
	f	9	5	17	10	21	26	9	20	2	1	25
	g	9	6	18	11	22	27	10	21	2	0	25
	A	10	0	19	12	23	28	11	22	1	6	25
	b	10	1	20	13	24	29	12	23	1	5	24
	c	10	2	21	14	25	30	13	24	1	4	24

Expliciũt table tablarum astronomice Raby abraham zacuti astronomi sereni[ssi]mi Regis emanuel Rex portugalie et cet cũ canonib⁹ traductis a linga ebrayca in latinũ p magistrũ Joseph vizinũ discipulũ ei⁹ actoris opera et arte viri solertis magistri ortas curaq[ue] sua nõ mediocri inprẽsione cóplete existũt felicib⁹ astris año a p[ri]ma rerum etherearum circuitione 1496 sole existẽte in 15 g 53 m 35 s piscium sub celo leyree

Figure 1: Colophon, *Almanach Perpetuum*, Leiria, 1496, f. 172r (Madrid, Biblioteca Nacional, shelfmark I-1077).

or on board a ship prior to his arrival in Portugal. Moreover, he does not discuss astronomical instruments in any of his known works, and he does not mention the application of astronomy to problems of navigation. In fact, he only mentions four astronomical observations that he made, and gives no details about what instruments he used, if any: (1) a remark on an occultation of Spica by the Moon in 1474 (Cantera 1931, p. 194); (2) a detailed discussion of the occultation of Venus by the Moon on July 24, 1476 (Goldstein and Chabás 1999); (3) a brief report of the total solar eclipse of July 29, 1478 (Zacut, *Sefer Yuḥasin*, p. 226b); and (4) a notice of the solar eclipse of March 16, 1485 (Carvalho 1927, p. 40). So, while we cannot deny the possibility of some interaction between Zacut and navigators and explorers, we await better information that is closer in time to Zacut's sojourn in Portugal than is provided by Correia.

Zacut's treatment of the occultation of Venus is particularly noteworthy because he considers modifications of the Ptolemaic models for planetary latitude based on his own observation, whereas in general he has simply depended uncritically on traditional models (see comments on Table AP 37, below). Elsewhere in medieval astronomy there is rarely any discussion of latitude theory other than presentations of either Ptolemy's models or those that derive from Hindu sources. Despite its unusual character, Zacut's report of the occultation of Venus is only preserved in one Hebrew manuscript (MS B), and is not mentioned in any Latin source of which we are aware.

The document, mentioned above, concerning a payment to Zacut by the Portuguese government, supposedly contains his signature, but the authenticity of that signature has been called into question on several grounds by D. Kaufmann (1897). The signature begins with an abbreviation for "rabbi"—this was contrary to the usual custom at the time when it was considered boastful to designate oneself in Hebrew in that way. The title that comes after the name "Abraham Zacut," "astronomer of the king (*ha-tokhen meha-melekh*), João [?]," looks like a literal translation from the Portuguese (or possibly Castilian), rather than an expression in idiomatic Hebrew that one would expect from a major Hebrew author. And Kaufmann noted that the facsimile version provided by Kayserling (1896) has some letters in this title crossed out to improve the Hebrew style, which is very odd in a signature. But in a facsimile recently provided to us by the director of the Portuguese archives there are no letters crossed out, and this suggests that the facsimile published by Kayserling had been altered. No other signature by Zacut has been found with which to compare it.

We have shown Figure 2 to a number of specialists in Hebrew palaeography in Jerusalem, including E. Engel, M. Fraenkel, Y. T. Langermann, and B. Richler, and they have informed us that

1. It is unusual, especially if the document is really a receipt, for the signature to appear so far below the text.

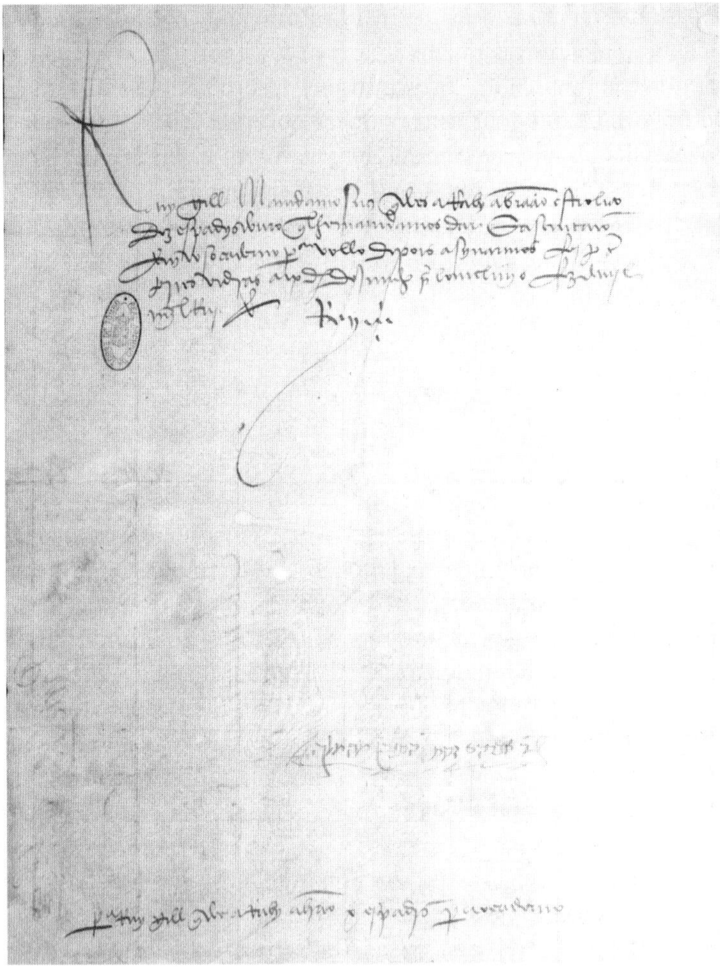

Figure 2: Document with Zacut's signature (Lisbon, Arquivo Nacional da Torre do Tombo, *Corpo Chronologico*, parte 1ª, maço 2, doc. 18).

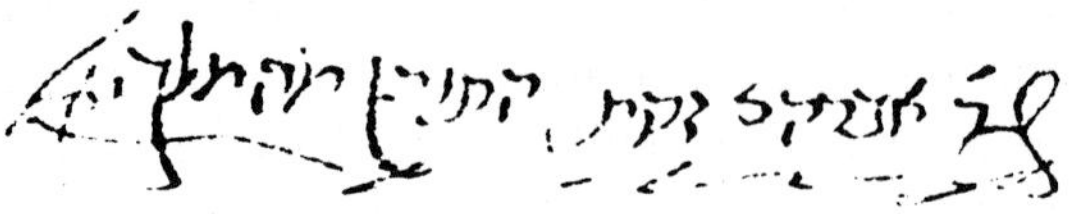

Detail of Figure 2.

2. It looks very much like the words *ha-tokhen meha-melekh* were added after the signature, possibly by a different person. They are not exactly on the same line, there are subtle differences betwen the letters, and there is even a sign of some sort of crease, fold, or other marking between the two sets of words.

3. The name "R. Abraham Zakkut" [= Zacut] is not written in a hand which is purely Sefardic, but displays other, Italian influences; E. Engel pointed to the unusual *zayin* and also the way the *kaf* and *waw* are connected in the name Zakkut.

4. M. Fraenkel can only confirm the unlikelihood that Zakkut would sign "Rabbi," and has not seen anything like the flourish before the name. She also pointed out that the flourish seems to be related to the odd way that the name is underlined.

During Zacut's stay in Portugal (ca. 1492–1497), a high ranking ecclesiastical dignitary at the Portuguese Court, formerly professor of astronomy in Salamanca, may have played an important role in providing Zacut with access to the royal court: Diego Ortiz de Calçadilla.

Diego had succeeded Juan de Salaya in 1469 in the chair of astronomy at the University of Salamanca, but had held various academic positions at least since 1465 (see Cantera 1931, pp. 373–376). There is no evidence that Zacut ever met Ortiz in Salamanca, although it is difficult to exclude the possibility since Zacut was 24 years old when Ortiz left Salamanca definitively. On the other hand, some documents concerning the University of Salamanca show that Ortiz was well acquainted with Salaya, with whom Zacut had some contact. Indeed, Beaujouan (1966, p. 77) suggested that there was a close relationship between Zacut and Ortiz as well as between Zacut and Salaya. To be sure, Beaujouan did not claim that there was direct evidence linking Zacut and Ortiz and, after a review of the known documents, we have not found any.

Ortiz left Salamanca for Portugal in 1476 (Lucena e Vale 1934, pp. 27 ff). After the death in 1474 of the King of Castile, Enrique IV, his sister Isabel and his daughter Juana disputed the succession to the throne. Isabel was supported by the Castilian nobility, and proclaimed herself Queen of Castile. Juana, called "la Beltraneja," was supported by Portugal whose king, Afonso V, had married her when she was 12 years old. The battle of Toro (1476) put an end to the dispute in favor of Isabel. Princess Juana's followers, and among them Diego Ortiz—her confessor, according to Lucena e Vale—had to leave for Portugal. Shortly thereafter Ortiz's chair at the University of Salamanca was declared vacant.

Diogo Ortiz de Vilhegas was "Castilian by nation," as attested in various Portuguese documents (see Góis 1690, I:131–132, cited in Almeida 1967, I:432 ff), and we accept the claim that he is identical with Diego

Ortiz, who had been professor of astronomy in Salamanca. He was from Calçadilla, a village that Lucena e Vale locates in the Castilian province of León (p. 27), where his parents, Iñigo Ortiz and Isabel de Villegas, lived. His ecclesiastical career in Portugal was certainly successful, for he became bishop of Tangier (1491–1500), Ceuta (1500–1505), and Viseu (1505–1519), and held many other prominent posts within the structure of the Church in Portugal. He is said to be the author of some works of religious content (Lucena e Vale 1934, pp. 207–208). Diogo Ortiz was also known as a preacher, and he was asked to deliver sermons at the official ceremonies organized for some navigational exploits: the return of Vasco da Gama from India (1499), the departure of Pedro Alvares Cabral's fleet to Brazil (1500), and the return of Duarte Pacheco from the orient (1505).

Diogo Ortiz de Vilhegas is not to be confused with his two nephews bearing the same name. Just to add some more confusion to the story, one of them also served as bishop of Ceuta (1540).

There are indeed quite a few extant documents showing Diogo Ortiz as a man of the Church who had the confidence of the successive kings of Portugal: Afonso V, João II, and Manuel I, but there are only two where Ortiz is in one way or another related to scientific activity. One concerns his participation in a group to counsel the king on navigation, and the other refers to his contribution in the making of a map.

Diogo Ortiz was supposedly the head of the so called "Junta dos Mathematicos" that gave advice to King João of Portugal (reigned: 1481–1495) concerning navigation, and that rejected proposals of Christopher Columbus in 1482. The literature on this "Junta" is indeed confusing and not very convincing. For the historian and chronicler João de Barros (ca. 1496–1570), this group of advisers was formed by "D. Diogo Ortiz Bispo de Cepta, e com Mestre Rodrigo, e Mestre Josepe." The expression, "Junta dos Mathematicos," has also been used to designate a different group of royal counselors appearing later in de Barros's account, when he explains that King João proposed to "Mestre Rodrigo, e a Mestre Josepe Judeo, ambos seus medicos, e a hum Martim de Boemia" to address the problem of navigation at sea by means of the solar altitude. The expression "Junta dos Mathematicos" of which Diogo is supposed to have been a member seems to be a modern expression that is not found in de Barros (at least it is not found in the passages quoted by Bensaude 1912, pp. 260–264; see also Maddison 1992, pp. 97–98).

According to Lucena e Vale (1934, p. 69), who depended on P. Francisco Alvares (*Verdadeira Informação das Terras de Preste João das Indias*, chap. 103), before sailing to India in 1487, Pero da Covilhã and Afonso de Paiva were given a map drawn up by Diogo Ortiz and others: "huma carta de marear tirada do mapamundo & q. forã aho fazer desta ho licenceado Calçadilha q. agora he bpo de Viseu, & o doutor mestre Rodrigo morador ás Pedras negras & o doutor mestre Moysés a este tempo judeu [. . .] & el-rei lhe dera para ambos c.c.c.c. cruzados" (a sailing map taken from the world map and made (?) by master Calçadilha, now Bishop of Viseu, and master Rodrigo, living in Pedras Negras, and master Moysés, still a Jew at the time [. . .] and the King gave them both 400 *cruzados*).

Diego Ortiz de Calçadilla (and later de Vilhegas) may have played an important role in Zacut's activity in Portugal, possibly introducing him to the royal court, but we have not found any trace of this. Moreover, Diego's interest in astronomy (rather than geography and navigation) is not in evidence in the 43 years he spent in Portugal after leaving his chair of astronomy in Salamanca until his death in 1519.

At the time when the practice of Judaism was declared illegal in Portugal, Zacut traveled to North Africa, and settled in Tunis, where he wrote *Sefer Yuḥasin*. In this book we read that he and his son, Samuel, came to Africa and were imprisoned on two occasions (p. 223a). While in North Africa, in 1498 he also composed a text in which he applied an astrological theory of history according to which eclipses and planetary conjunctions are used to determine the date of messianic fulfillment. On the basis of this theory, he believed that the salvation of Israel would begin in 1503/4.[9] Zacut continued his astronomical activity and adapted the tables of the *Ḥibbur* for the year 1501. Later, he compiled another set of astronomical tables, beginning with the year 1513, arranged for the Jewish calendar and the meridian of Jerusalem, where he then lived.[10] Zacut is also the author of a short astrological work entitled *Mishpeṭei ha-'iṣṭagnin* (*Judgments of the astrologer*) concerning the years 1518–1524, which was transcribed by C. Roth (1949).[11] According to Roth, Zacut died in 1515, probably in Damascus. It has recently been argued by A. David, however, that Zacut died in Jerusalem.[12]

---

[9] See Goldstein (1998); Beit-Arié and Idel (1979).

[10] For Zacut's sojourn in the eastern Mediterranean area and the tables he compiled there, see Goldstein (1981).

[11] See especially pp. 447–448. A translation is offered in Appendix 1.

[12] David (1992, p. 81) wrote concerning Abraham ha-Levi that his brother-in-law, "the famous astronomer and historian R. Abraham Zacuto arrived in Jerusalem in 1513 and studied in one of the *yeshivot* (academies) there. He died in Jerusalem a year later." Abraham ha-Levi is known for his apocalyptic and kabbalistic works. According to David (1992, pp. 89–90), "he believed that redemption was imminent, expecting it to occur in stages at certain times: in 1520, 1524 and 1529. He also believed that the Messiah would appear in Safed. His apocalyptic visions were undoubtedly supported by the astrological calculations determined in the tables drawn up by his brother-in-law, R. Abraham Zacuto. Abraham ha-Levi died between 1530 and 1535."

2. Setting the Scene

2.1 The almanac tradition

In the medieval astronomical literature the term almanac was used to designate various kinds of texts. Even the etymology of this term is subject to dispute, and its ultimate origin remains unknown (see, e.g., Benjamin and Toomer 1971, p. 375). We shall consider here that an almanac is a set of tables giving the daily (or at intervals of a few days) true positions in longitude of the Sun, the Moon, and the five planets for a period of recurrence that is different for each celestial body. This kind of almanac was certainly more "user-friendly" than the standard sets of tables that only give tabulated data for mean motions and equations to compute true positions in longitude. But an almanac-maker had to face a real challenge because of the enormous number of time-consuming calculations. Nevertheless, medieval astronomers managed to avoid some of the calculations by reaching a compromise between accuracy and computational effort.

The periods of recurrence for planetary positions were already known to Babylonian astronomers. They are preserved in the so-called "Goal-year texts" (Neugebauer 1975, p. 554) and were used for predicting lunar and planetary phenomena. By some mode of transmission that is not fully understood, these goal-year periods reached Greek astronomers in late antiquity. In the *Almagest* (IX, 3), Ptolemy gives the following definition for such periods: "the smallest period in which it [each of the planets] makes an approximate return in both anomalies." The problem consists in finding for each planet an integer number of solar years (N) spanning the same time as an integer number of revolutions in longitude (R), and corresponding to an integer number of returns in anomaly (A) (Neugebauer 1975, pp. 150–151). The values given by Ptolemy are the following:

Planet	N	R	A
Saturn	59y + 1;45d	2 rev. + 1;43°	57
Jupiter	71y – 4;54d	6 rev. – 4;50°	65
Mars	79y + 3;13d	42 rev. + 3;10°	37
Venus	8y – 2;18d	8 rev. – 2;15°	5
Mercury	46y + 1; 2d	46 rev. + 1°	145

It is readily seen that N = R + A for the outer planets whereas N = R for the inner planets.

The earliest known almanac compiled in Muslim Spain is the Almanac of Azarquiel (Millás 1943–1950, pp. 72–237). It contains many tables in addition to those giving the positions of the five planets and the two luminaries. In this almanac, the epoch is September 1, 1088 (beginning of year 1400, Era of Alexander), and the tables are based on almost the same periods as those of Ptolemy.

Saturn	59y + 1;55°
Jupiter	83y − 2°
Mars	79y + 1°
Venus	8y + 1;30°
Mercury	46y − 2;45°

Note that the period for Jupiter differs from that in the *Almagest,* but agrees with another value for Jupiter (83y) that is attested in the "Goal-year texts" (for Mars 47y is attested as well).

The Almanac of Azarquiel gives true positions for the Sun and the planets, whereas it just gives mean positions for the Moon (longitude, anomaly, and lunar node). The planetary positions are given in sidereal coordinates. Table 1, below, displays the basic information on the periods, the frequency, and the accuracy of the entries in the Almanac of Azarquiel.

Azarquiel's approach was followed by several later astronomers. Millás (1946) described and attributed to R. Abraham Ibn ʿEzra (c. 1089–1167) a short text explaining an almanac, and extant only in Latin (London, British Library, MS Vesp. F11; Erfurt, Bibl. Amploniana, MS Q381). This text has not yet been adequately studied, and no tables are associated with it. Ibn ʿEzra certainly knew about almanacs, for in his *De rationibus tabularum,* a text extant in Latin only, we find the earliest occurrence of the term "almanac": "Sed vere sunt tabule que singulis diebus docent coequare planetas vel a tempore determinato dant rationes componendi almanac, id est tabulas per quas semel factas per totum annum planetas coequatos habebis" (Millás 1947, p. 119:2–23).

Ibn al-Bannā' (1256–1321), an astronomer of Marrakesh (although of Hispano–Muslim origin), seems to be the author of another almanac in the tradition of Azarquiel. It is extant in British Library, MS Arabic 977, and some references to it can be found in Millás (1943–1950, pp. 348 ff). Samsó (1997, p. 74) has recently called attention to another Arabic almanac in the same tradition and compiled by an anonymous astronomer (ca. 1266–1281).

Almanacs are an astronomical genre not exclusively developed in al-Andalus and the Maghrib. In Southern France Jacob ben Makhir (Prophatius) composed an almanac in Hebrew for Montpellier with March 1, 1301 as epoch. Boffito and Melzi d'Eril (1908) published a Latin version of it based on a manuscript in Florence. The tabulated values for the positions of the Sun begin at that date, but those for the planetary positions, given in tropical coordinates, begin in March 1300. It has been shown that all entries rely on the Toledan Tables (Toomer 1973); their main features appear in Table 1.

The so-called "Almanac of Tortosa" has been preserved in Latin, Catalan, Castilian, and Portuguese, but not in its original Arabic version. The identification of its city of origin as Tortosa was made when only one version, copied in that Mediterranean Spanish city, was known. It is thus preferable to call it "The Almanac of 1307," for the tables associated with it have March 1, 1307 as epoch. The tables have been published by Rico y Sinobas (1866), in the same volume as his version of the *Libro de las tablas alfonsíes*,[1] and he claimed that they were "numerical fragments" of the original Alfonsine Tables. Again, Table 1 shows the main characteristics of the Almanac of 1307. This almanac must have been quite popular, given the numerous copies of it in different languages. There were also later adaptations of it, such as that due to Ferrand Martines in the late fourteenth century (Chabás 1996b).

To be sure, there were other almanacs that were compiled outside the Iberian Peninsula and its area of influence. In particular, at the end of the thirteenth century and the beginning of the fourteenth century in Paris William of Saint-Cloud, John of Lignères, and John of Saxony compiled tables in almanac form, some of them based in the Alfonsine Tables, but it is unlikely that any of them was the basis for Zacut's almanac.

2.2 The Alfonsine Tables and Salamanca

IT HAS BEEN SAID that European Latin astronomy from 1320 to the sixteenth century is virtually co-extensive with the history of the Alfonsine Tables (North 1977, p. 270). While this statement can easily be supported with respect to most of Europe, it clearly does not apply to Spain, the very place where the Alfonsine Tables were produced in the thirteenth century. Indeed, there is no evidence that these tables were used before 1460 by anyone in Spain despite a number of allusions to them, as is the case, for example, in the canons to the Tables of Barcelona (Millás 1962, p. 90:6–7), and in the canons to Jacob ben David Bonjorn's tables (Chabás 1992, pp. 180–181:27). The canons to the Alfonsine Tables survive in the original Castilian, but the tables that belong with them are not extant. In fact, the diffusion of these tables came mainly from Paris where a revision was undertaken by John of Lignères and John of Murs in the 1320s, often

[1] In his version of the unique manuscript containing this book, Rico has *taulas* instead of *tablas* (see Madrid, Biblioteca Nacional, MS 3306, f. 35r).

Table 1: Comparison of the entries in several almanacs. In the following table, *r* stands for the length of the period under consideration, *f* is the frequency of the entries, and *a* is the accuracy of the entries (two values for *f*, e.g., 1y and 1d, means that 2 tables are presented, one at intervals of 1 year, and one at intervals of 1 day).

	Azarquiel	**J. ben Makhir**	**Almanac of 1307**
Sun: longitude	r = 4y f = 1d a = 0;1°	r = 4y f = 1d a = 0;0,1°	r = 4y f = 1d a = 0;0,1°
Moon: longitude	r = 76y f = 1y, 1d a = 0;1°	r = 76y f = 1y, 1d a = 0;0,1°	r = 76y f = 1y, 1d a = 0;0,1°
Moon: anomaly	r = 180y f = 1y, 1d a = 1°	r = 24y f = 1 d a = 0;1°	r = 86y f = 1y, 1d a = 0;1°
Lunar node	r = 93y f = 1y, 5d a = 0;1°	r = 93y f = 1y, 1d a = 0;0,1°	r = 93y f = 1y, 3d a = 0;1°
Saturn: longitude	r = 59y f = 10d (1) a = 1°	r = 59y f = 10d (3) a = 0;1°	r = 59y f = 10d (1) a = 1°
Jupiter: longitude	r = 83y f = 10d (1) a = 1°	r = 83y f = 10d (3) a = 0;1°	r = 83y f = 10d (1) a = 1°
Mars: longitude	r = 79y f = 5d (2) a = 1°	r = 79y f = 10d (3) a = 0;1°	r = 79y f = 5d (2) a = 1°
Venus: longitude	r = 8y f = 5d (2) a = 1°	r = 8y f = 5d (4) a = 0;1°	r = 8y f = 5d (2) a = 1°
Mercury: longitude	r = 46y f = 5d (2) a = 1°	r = 46y f = 5d (4) a = 0;1°	r = 46y f = 5d (2) a = 1°

(1) Entries for days 1, 11, and 21 of each month.
(2) Entries for days 1, 6, 11, 16, 21, and 26 of each month.
(3) Entries for days 10, 20, and the last of each month.
(4) Entries for days 5, 10, 15, 20, 25, and the last of each month.

accompanied by the canons composed by John of Saxony in 1327, and this version served as the basis for the *editio princeps* (Ratdolt 1483).

In this Section we shall discuss the first evidence for the use of the Alfonsine Tables in Spain, the availability of the Alfonsine Tables in Hebrew, and Zacut's relationship to the traditions surrounding these tables.

Of special importance in this context is a text written by Nicholaus Polonius for his students at the University of Salamanca where he was the first to hold the chair of astronomy/astrology *ca.* 1460. This text is preserved in a complex Latin manuscript of Spanish origin (Oxford, Bodleian Library, MS Can. Misc. 27) that is of fundamental significance for the history of astronomy in Spain. Although it contains many astronomical texts and tables in disorder, generally with no indication of their beginning and end, we have identified the following items in it: the canons and tables of Jacob ben David Bonjorn (ff. 1r-31v); a set of tables that come from *al-Zīj al-Muqtabis* of Ibn al-Kammād (ff. 110r-111r); some tables attributed to John of Lignères (beginning on f. 88v); some isolated canons of the "Priores astrologi motus corporum" by the same author: chapters 19–24, 31, 33–39, 41–46 (ff. 130r-138r), and all 44 chapters of another text of John of Lignères beginning "Cujuslibet arcus propositi sinum" (ff. 138r-145v).

Polonius's text is found on ff. 122v-129r, and begins with "Quoniam tabularum Alfonsi laboriosa difficultas"; for a transcription of the text, see Porres and Chabás (1998). It is arranged in 18 chapters, and deals with some astronomical tables for Salamanca: the *Tabulae Resolutae*, also extant in the same manuscript (Chabás 1998). The introduction indicates that these canons were intended for students in Salamanca, and explains, among other matters, how to compute the positions of the Sun, the Moon, the 5 planets, ascendants, and other astrological magnitudes, times of syzygies, and how to compile an almanac (chapter 18, entitled "Almanach pro singulis annis componere"). Polonius's canons to the *Tabulae Resolutae* were adapted from those by Andreas Grzymała of Poznan for his students at the University of Cracow (1449),[2] and are in turn largely based on the canons to the Alfonsine Tables composed in Paris by John of Saxony.

The tables associated with these canons are on ff. 33r-88r. The first table has the heading "Tabule ad meridianum Salamantinum," and the tables that follow are indeed calculated for Salamanca with year 1460 as epoch. They use zodiacal signs of 30°, the years are numbered as "completed years" (i.e., one less than the "current year"), and begin in January (i.e., noon of December 31). This set of tables is based on the Alfonsine Tables, and in fact it represents the earliest evidence so far for the use of them in the Iberian Peninsula after their compilation by the astronomers in the service of King Alfonso X of Castile. These tables for Salamanca follow very closely a particular form of presenting the Alfonsine Tables, called *Tabulae Resolutae,* extant in many fifteenth century copies from Central Europe,

[2] See Porres and Chabás (1998). The incipit of Grzymała's canons is an acronym of the author's name: "Gyrum recensendo zodiaci inter magnalia astronomie longe alto ingenio . . . ." These canons have never been edited, and we have consulted copies in Cracow, Jagiellonian Library, MSS 573, 1864, and 1865.

especially Poland (Dobrzycki 1987). This form of presentation differs in many ways from that in the *editio princeps* of the Alfonsine Tables.

The first 12 tables (ff. 33r-38v) give the mean motions of 12 quantities: longitude of apogees and fixed stars; access and recess; longitude of the Sun, Venus, and Mercury; longitude of the Moon; argument of the Moon; argument of lunar latitude; longitude of the lunar node; longitude of Mars; longitude of Jupiter; longitude of Saturn; argument of Venus; and argument of Mercury. They are displayed according to a system of cyclical radices "ad annos collectos" every 20 years, covering the period 1348–1628. The mean motions for the 12 quantities are given for 20 consecutive years ("anni expansi"), for months of the year (both for leap years and common years), days and hours. We are also given the mean motions for accumulated periods of 40, 60, 80, 100, 200, . . . , 1000 years (ff. 39r-40r). All entries are given to sexagesimal thirds, and all parameters for mean motions used here are the same as those in the 1483 edition of the Alfonsine Tables. It is worth noting that according to chapter 15 of the Castilian canons of the *Libro de las tablas alfonsíes* the original Alfonsine tables had their mean motions organized in 20-year periods, with signs of 30°, which is exactly the case in our manuscript, but not that of the Parisian version of the Alfonsine Tables. The *Tabulae Resolutae* for Salamanca also have tables not included in the 1483 edition, and tables based on different parameters, but found in pre-Alfonsine al-Andalus.

The *Tabulae Resolutae* for Salamanca, consisting of canons and tables adapted by Nicholaus Polonius for the city at whose university he was teaching, seem to be the earliest evidence for the use of the Alfonsine Tables in the Iberian Peninsula, and date from ca. 1460. In the decades following the introduction of the *Tabulae Resolutae*, intensive astronomical activity took place in Salamanca, as attested by various Latin and Castilian manuscripts containing astronomical material specifically for that city which will be examined below (see chapters 2.3 to 2.5), among them, Madrid, Biblioteca Nacional, MS 3385. This fifteenth century codex is of great interest because it contains a variety of tables and texts on astronomical matters, some of which have only survived in it. In addition to most of the tables of the *Ḥibbur*, this manuscript contains another set of tables for Salamanca (see chap. 2.3), the canons and the tables of Jacob ben David Bonjorn (1361: see chap. 2.6), a set of tables with their headings in Castilian (see chap. 2.4), and some treatises by Diego de Torres (see chap. 5.4), indicating the close relationship between Zacut and Christian scholars in Salamanca (although this is not mentioned either by Zacut or by these Christian scholars).

As far as we can tell, there were five Hebrew versions of the Alfonsine Tables, of which 3 are close to the 'standard' form in the *editio princeps*, and 2 are versions of Batecomb's 1348 tables for Oxford that are notable for including double argument tables for finding planetary longitudes (North

1977, pp. 279 ff). Of greatest interest for the study of Zacut is the version that, in our view, was intended for Salamanca (No. 5, below); the others were not produced in Spain, and there is no direct evidence to indicate that Zacut was aware of them (but see the discussion of Tables AP 23, AP 27, AP 32, AP 37, AP 43, in chap. 4.3). We will treat these versions in chronological order.

1. Solomon ben Davin de Rodez, a pupil of Bonfils (*ca.* 1350), produced a Hebrew version of Batecomb's 1348 tables for Oxford, but modified for 1368 and the meridian of Paris (Munich, MS Heb. 343, ff. 104b-167a). On f. 108a are radices for noon of the last day of December 1368, for the horizons (i.e., geographical latitudes) of Avignon, Paris, and Lyon. Another copy of this Hebrew version is preserved in Oxford, Bodleian Library, MS Regio 14, ff. 57a-103b (Goldstein 1979, p. 36).

2. In 1441 Mordecai Finzi of Mantua produced another Hebrew version of Batecomb's 1348 tables (Oxford, Bodleian Library, MS Lyell Heb. 96; Goldstein 1987, p. 120; and Langermann 1988, pp. 26–28).

3. In 1460 Moses ben Abraham de Nîmes in Avignon translated the Alfonsine Tables from Latin, including the canons by John of Saxony (Munich, MS Heb. 126). This version is very close to the *editio princeps* but, on f. 69b, it is said that 8 tables were added by later Christian scholars, and were not part of the original tables (Goldstein 1980, pp. 137–138).

4. An anonymous version, probably composed in Italy in the late fifteenth century, is preserved in a manuscript of the fifteenth or sixteenth century (Milan, Ambrosiana, MS Heb. X-193 Sup., ff. 1a-27b; Luzzatto 1972, p. 90, lists it as "unidentified tables"). The mean motions and equations agree with the *editio princeps*, but the radices (f. 1a) seem to be arranged for dates in the Jewish calendar to be converted to a number of days in sexagesimal form. In this manuscript sexagesimal digits are to be read from right to left (as is usual in Hebrew astronomical tables), but each sexagesimal digit is written decimally, to be read from left to right, using the first nine letters of the Hebrew alphabet together with a symbol for zero. In the equation tables, the arguments are given in signs of 30° and not in signs of 60°, as in the *editio princeps*. This mixed system of notation is very unusual, and possibly unique to this manuscript. In another part of this manuscript (f. 48b), there is a table, entitled: "Sunrise in Parma according to Joanni Bianchino," i.e., Giovanni Bianchini (d. after 1469).

5. Finally, we come to a Hebrew version of the Alfonsine Tables that is almost certainly related to Salamanca (St. Petersburg, Academy of Sciences, MS Heb. C-076, ff. 31a-54a). On f. 32b are two lists of radices, one for epoch Alfonso, May 31, 1252 "for longitude 28;30°" (i.e., Toledo), and another for the beginning of 1473 (place unspecified). These tables are followed by an anonymous commentary on the Alfonsine Tables (ff. 55a-56b), and in it two epochs are mentioned, "the last day of May 1252," and

the beginning of 1473, "that is, noon of the last day of December 1472" (f. 55b); hence this commentary is related to the tables that precede it. It may be of interest that the epoch of Alfonso here agrees with the Parisian version of the Alfonsine Tables, whereas in the Castilian canons the epoch of Alfonso is taken to be noon preceding Jan. 1, 1252. Although the commentary does not indicate the place for which the tables were arranged, we are confident that the place was Salamanca. Our reasoning is that there are 80569 days from the epoch of Alfonso to noon of Dec. 31, 1472 and, with the Alfonsine parameters, this yields the radices for the planets listed on f. 32b for the later epoch. But for the Moon, we used a more precise value for the time from the epoch of Alfonso (Toledo) to noon of Dec. 31, 1472 (Salamanca): 80569;0,27,20d, and this yielded much better results. The difference in longitude between Toledo and Salamanca is 2;44° according to Table TV 19, or 0;10,56h which is 0;0,27,20d. In the manuscript, the mean position of the Moon at noon on Dec. 31, 1472 is 305;7,59,0°. Recomputing this longitude with 80569;0,27,20 days after the radix for Alfonso yields 305;7,58,48° that differs from the text by only 0;0,0,12°, whereas recomputing with 80569 days yields 305;1,58,39° that differs from the text by about 0;6°. Since Zacut's tables are arranged for 1473, it seems most likely that this version of the Alfonsine Tables has something to do with the community of scholars in Salamanca at the time of Zacut.

As we shall see below (chap. 2.6), Zacut does not acknowledge his Christian predecessors or contemporaries in Salamanca but, on the other hand, he cites a large number of Jewish astronomers. Nevertheless, in the course of our research it became clear that Zacut was heavily, and perhaps primarily, dependent on this Christian context, and that the Jewish tradition in astronomy played a subordinate role. To be sure, some Jewish elements had already been incorporated into the study of astronomy by Christian scholars, but the essential ingredient was the introduction of a Latin version of the Alfonsine Tables brought to Salamanca in the 1460s.

2.3 The *Tabule Verificate* for Salamanca

IN MADRID, BIBLIOTECA NACIONAL, MS 3385 (ff. 104r-113r), there is a set of 21 tables with headings in Latin for computing eclipses, not previously studied, that we will call *Tabule Verificate* for Salamanca. The general heading for these tables is: *Incipiunt Tabule verificate de o–[conjunctionibus] et o–o [oppositionibus] solis et lune reducte ad meridianum Salamantinum.* Each of the tables will be preceded by the *siglum* "TV."

These tables use signs of 30 degrees, with a year beginning in January except for Tables TV 10 and TV 11 (solar elongation from the lunar node, and motion of the lunar nodes). The epoch for these tables is January 1, 1461, and they were calculated for the city of Salamanca. This indicates that they were composed at a time close to that date. It would seem unlikely for anyone to go through so much effort in computing many of the entries in the mean motion tables at a date after, say, 1464. The only author we can suggest is Polonius himself. Salaya might be considered as another possible candidate, but his translation of the *Ḥibbur* indicates that he did not completely master Zacut's text. If Polonius is the author, there is some difficulty in explaining the fact that the *Tabule Verificate* contain tables from Bonjorn (Tables TV 15 and TV 16). What is certain is that the author made a good selection of the material at hand and gathered some of the best tables available. He knew what he was doing!

The first 11 tables consist of Alfonsine material adapted to Salamanca. In addition to the standard Alfonsine material found, for instance, in Polonius's *Tabulae Resolutae* for Salamanca, there is also a table for solar and lunar corrections by Nicholaus de Heybech. It is likely that Polonius brought it with him from Poland, for numerous manuscripts in Cracow attest that Heybech's table was widely available in Poland at the time. There can be no doubt that Zacut used this material: in fact, he reproduced or recast Tables TV 9 to TV 17, and put in words Table TV 18. Clearly this is a primary source for Zacut.

TV 1. Table for the elongation (*sic*) at mean conjunction (f. 104r)

According to its title, this tables displays the "elongation at mean conjunction at the beginning of 76 years." The tabulated values are given in days, hours, and minutes, for each year in a period of 76 years. The value for year 77 is also given. The unit used indicates that this "elongation" is not to be considered as an angular distance but as time. Actually, the tabulated magnitude is the period of time between the beginning of each year and the first mean conjunction of the Sun and the Moon. The entry for year 1 is 18d 8;49h.

The entries in this table can be recomputed under the assumption that the table was calculated for the longitude of Salamanca for January 1, 1461 as epoch, and that it is based on the Alfonsine Tables. It has been shown that a version of the Alfonsine Tables was available at Salamanca no later than 1460: the *Tabulae Resolutae* adapted for this Castilian city by Nicholaus Polonius (or Nicolás Polonio, as he is called in Castilian), probably of Polish origin, and extant in Oxford, Bodleian Library, MS Can. Misc. 27, ff. 33r-88r (Chabás 1998; Porres and Chabás 1998).

The mean solar position at noon on January 1, 1461 can be computed from Polonius's tables for Salamanca by adding the corresponding entries for 1448 *anni collecti* (9s 18;59,45,34°) and 12 *anni expansi* (0s 0;5,17,24°), yielding 9s 19;5,2,58°. Note that Polonius's tables also use signs of 30°.

The mean lunar position is obtained by adding the corresponding entries in Polonius's tables for 1448 *anni collecti* (0s 0;51,25,56°) and 12 *anni expansi* (5s 2;8,16,32°), yielding 5s 2;59,42,28°. The resulting elongation is 7s 13;54,39,30°. Dividing this number by the Alfonsine value for the mean motion in elongation in Polonius's tables (13;10,35,1 – 0;59,8,20° = 12;11,26,41°/d), we obtain the time from the beginning of 1461 to the first mean conjunction: 18d 8;48,54h. This is in full agreement with the entry in the table, and thus validates our assumptions. Of course, the same result to minutes could have been obtained using any other good set of Alfonsine Tables, and taking into account the difference in longitude between Toledo (28;30°) and Salamanca (25;46°), as reported in various manuscripts, e.g. Oxford, Bodleian Library, MS Can. Misc. 27, f. 122v; Madrid, Biblioteca Nacional, MS 3385, f. 112r.

TV 2. Tables for syzygies (f. 104r)

The first table displays, for each month of the year (both for a leap year and for a common year), the time (in days, hours, and minutes) needed to complete a full civil month after consecutive mean synodic months have elapsed, using the standard Alfonsine value, 29d 12;44h. The corresponding entry for January is 1d 11;16h, and it is computed by subtracting 29d 12;44h from 31d. The entry for February in a leap year is 0d 22;32h, and it is obtained subtracting twice the length of the mean synodic month from 60 days, and so on.

The second table displays multiples and sub-multiples of the mean synodic month useful in computing the moments of quadratures. The first entry is 7d 9;11h, which is exactly a fourth of the mean synodic month, mentioned above.

TV 3. Tables for the weekday at the beginning of the year (f. 104r)

In these two small tables we are given the weekday for the first day of the year for each year in a period of 28 years, and then for periods of 28 years. The entry for the first year is 7, Saturday.

TV 4. Table for the deficit with respect to an integer number of days (f. 104v)

For each integer between 1 and 26 in the first column, the second column gives the deficit with respect to an integer number of days between 2 syzygies at a distance of a multiple of 76 Julian years. The first entry is 0d 5;52h. The first 20 entries are multiples of the first entry for each integer number of cycles from 1 to 20, where the entry for 20 cycles is 4d

21;20h, but the 6 last tabulated values, intended for 40, 60, 80, 100, 120, and 140 cycles, are not correct.

The entry for year 1 is the value rounded from 5;52,13h, which is obtained by subtracting 940 mean synodic months of 29d 12;44,3,3h from 76 Julian years (27,759 days). This value for the mean synodic month is found in the Alfonsine corpus; it appears, for instance, in Polonius's *Tabulae Resolutae* for Salamanca (Oxford, Bodleian Library, MS Can. Misc. 27, f. 42v). Note that the "standard" value for the mean synodic month is that given in the Greek version of the *Almagest* (IV, 2): 29;31,50,8,20d (= 29d 12;44,3,20h) (Toomer 1984, p. 176).

TV 5. Table for the mean motion in lunar anomaly (ff. 104v-105r)

This table consists of a set of sub-tables that display the mean motion in lunar anomaly to minutes of arc for each year in a period of 180 years, for each month of the year (both for a leap year and for a common year), and for each day, hour, and minute. The value for year 181 is also given. Moreover, another sub-table lists the entries for periods of 180 years, where the entry for 1 period is 2;20°. As far as we know, 180 years as a period for lunar anomaly was used previously only once in a table of this kind, namely, in the Almanac of Azarquiel (Millás 1943–1950, p. 170).

In the main table, the entry for year 1 is 7s 10;10°, whereas that for year 181 is 7s 7;50°; the difference is 2;20°, in accordance with the tabulated value in the sub-table mentioned above. Madrid, MS 3385, reads 10s 8;43° for year 2 instead of 10s 8;53°, and 10s 7;36° for year 3 instead of 0s 7;36°.

The tabulated value for year 1 indicates that, as was the case with the table for the "elongation at mean conjunction," it was calculated for the longitude of Salamanca for January 1, 1461 as epoch, using the Alfonsine Tables. In Polonius's *Tabulae Resolutae* for Salamanca we find 6s 16;19,34,48° as the entry for 1448 *anni collecti* and 0s 23;50,39,48° for 12 *anni expansi*, yielding a total of 7s 10;10,14,36°, in full agreement with the tabulated value.

All other entries in the sub-tables for the mean motion in lunar anomaly rely on the basic Alfonsine parameter: 13;3,53,57,30,21°/d.

TV 6. Table for the mean motion in solar anomaly (f. 105v)

This table consists of a set of sub-tables that display the mean motion in solar anomaly to minutes of arc for each year in a period of 76 years, for each month of the year (both for a leap year and for a common year), and for each day, hour, and minute.

In the main table, the entry for year 1 is 6s 18;16°, thus indicating that, as was previously the case, the epoch used is January 1, 1461, and that the calculation was carried out from some set of Alfonsine Tables,

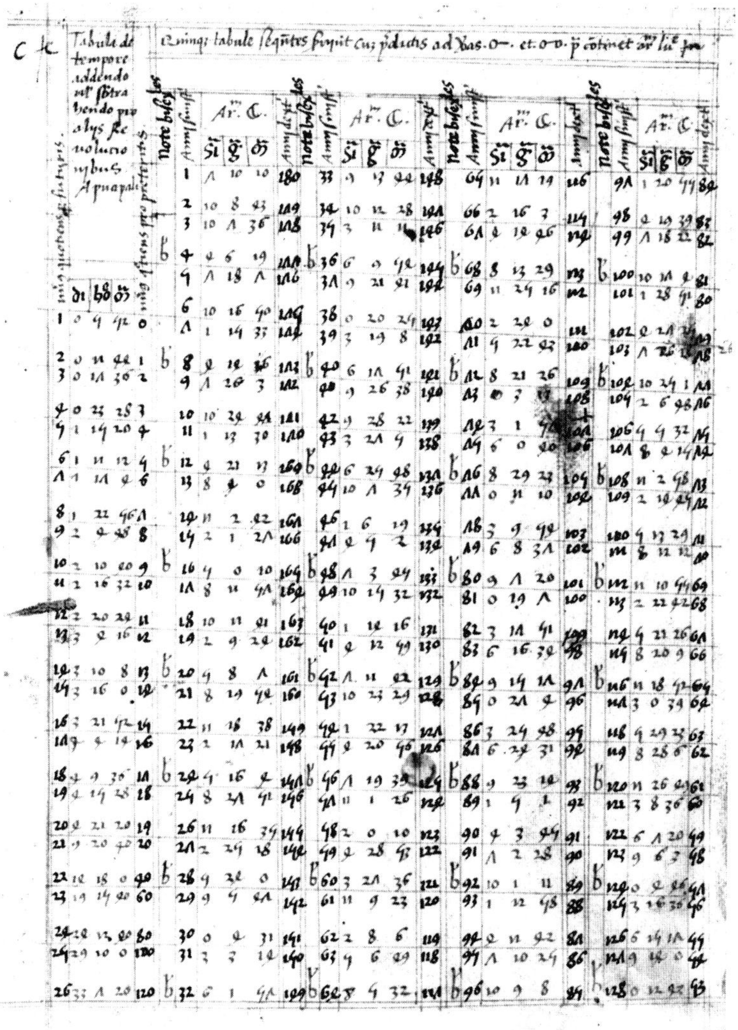

Figure 3. Table TV 4: Table for the deficit with respect to an integer number of days (Madrid, Biblioteca Nacional, MS 3385, f. 104v).

such as Polonius's tables. As noted above, the mean solar position for that date resulting from Polonius's *Tabulae Resolutae* for Salamanca is 9s 19;5,2,58°. Subtracting the position of the solar apogee given by Polonius (3s 0;48,38,17°: see Oxford, Bodleian Library, MS Can. Misc. 27, f. 39r) yields 6s 18;16,24,41°, also in full agreement with the tabulated value.

All other entries in the sub-tables for the mean motion in solar anomaly rely on the basic Alfonsine parameter: 0;59,8,19,37,19°/d.

TV 7. Table for the lunar correction (*equatio argumenti lune*) (f. 106r-v)

This table has 5 columns. Column 1 is the argument, given in integer degrees from 1° to 30° for each sign from 1 to 5 (i.e., from 1° to 180°). Column 2 gives the lunar correction in hours and minutes. Column 3 displays the line-by-line differences in the previous column, and only serves to facilitate interpolation. Column 4 lists the minutes of proportion, a monotonically decreasing function from 60 to 0. Column 5 shows the line-by-line differences in the previous column, and only serves to facilitate interpolation.

The columns for 3 signs are written twice, once on f. 106r and again on f. 106v. The maximum value for the lunar correction (col. 2) is 9;40h for all arguments from 84° to 90°.

The two significant columns (2 and 4) are taken from a table compiled by Nicholaus de Heybech of Erfurt (ca. 1400) to compute true syzygies (Chabás and Goldstein 1992). Nicholaus de Heybech's table is based exclusively on parameters found in the Alfonsine corpus, including a table for lunar velocity attributed to John of Genoa. It has survived in several Latin manuscripts, usually together with its accompanying canon. In 1992 we had located 4 copies of Heybech's work (2 in Paris, 1 in Dijon and 1 in Basel); we can now add to that list 7 other copies: 1 in Cusa (MS 211), 1 in Vienna (MS 2440), and 5 at the Jagiellonian Library, Cracow (MSS 609, 610, 613, and twice in 1865). Note that all these Cracow manuscripts contain material prior to 1460.

The ingenious solution suggested by Nicholaus de Heybech for determining true syzygies consists in dividing the time interval between mean and true syzygy into two terms, each of which accounts for the role of one luminary; then he treats each term separately, and within each term he computes a set of minimum and maximum values, and uses an interpolation scheme for intermediate values.

In particular, col. 2 is the same, but for copyist's errors, as Heybech's col. IV (see Chabás and Goldstein 1992, pp. 282–285), and its entries, c_2, can be computed as follows:

$$c_2 = c_m/[v_m(\bar{\alpha}) - \max(v_s)],$$

where c_m is the lunar equation, $\bar{\alpha}$ is the mean lunar anomaly, v_m is the lunar velocity and $\max(v_s)$ is the maximum solar velocity. The entries in col. 4, c_4, are obtained by subtracting from 60 min the corresponding entries in Heybech's col. III, and thus can be computed as follows:

$$c_4 = 60 - [(D - d(\kappa))/(D - d)],$$

where, in an eccentric model whose deferent has radius 60, D is the distance of the luminary at apogee from the observer, d its distance at perigee, and $d(\kappa)$ is its distance for a true anomaly κ.

TV 8. Table for the solar correction (*equatio centri vel argumenti solis*) (f. 107r-v)

This table has also 5 columns. Column 1 is the argument given in integer degrees from 1° to 30° for each degree sign from 1 to 5 (i.e., from 1° to 180°). Column 2 gives a correction for the Sun in hours and minutes. Column 3 displays the line-by-line differences in the previous column. Column 4 is labeled the "variation of the solar correction," in minutes. Column 5 shows the line-by-line differences in the previous column.

The maximum value for the solar correction (col. 2) is 3;46h for arguments from 89° to 96°. The maximum value for the "variation of the solar correction" (col. 4) is 61 min for arguments from 84° to 98°. The two significant columns (2 and 4) are taken from Heybech's table. In particular, col. 2 is the same, but for copyist's errors, as the difference between Heybech's col. I and II, and col. 4 is the same as Heybech's col. II (see Chabás and Goldstein 1992, pp. 282–285).

TV 9. Table for the equation of time (f. 108r)

This table lists the values, in minutes of time, of the equation of time, i.e., the difference between true and mean noon, for each day of the year, beginning in January (cf. Table AP 5). The extremal values are:

max = 0;23h:	30 April
min = 0;12h:	12–26 July
Max = 0;32h:	20–24 October
Min = 0; 0h:	22 January–6 February.

TV 10. Table for the solar elongation from the lunar node (ff. 108r-109r)

This table lists the differences in longitude between the Sun and the lunar ascending node for each year in a period of 56 years. Note that 56 years (= 28 · 2) correspond to two solar cycles. The entries are given in signs, degrees, and minutes. The value for year 57 is also indicated. This table is useful for a preliminary calculation of eclipse possibilities: when the elongation at syzygy is such that the Sun lies in the nodal zone, it is appropriate to compute the circumstances of a lunar or solar eclipse. There is a similar table, for the same purpose and the same argument, among the tables compiled by Ben Verga for Lisbon (epoch 1400), which uses a cycle of 28 years (see Table AP 12).

In the main table, the entry for year 1 is 8s 3;5°. Although it is not explicitly stated, this value corresponds to March 1, 1461. According to the Alfonsine Tables, the true position of the Sun at noon on the day before March 1, 1461 at Salamanca is 349;21,39°. At that moment, the longitude of the lunar ascending node is 106;17,22°, and the difference is 243;4,17°, in good agreement with the tabulated value (8s 3;5°). Note that, in contrast to all the preceding tables, this table begins in March.

A separate sub-table displays entries, also given in signs, degrees, and minutes, for each day in a year beginning on March 1 (the corresponding entry is 0s 1;3°). A more accurate value, resulting from the Alfonsine mean motions of the Sun and the lunar node, would be 0s 1;2,19°. Other sub-tables show entries, also given in signs, degrees, and minutes, for each hour in a day, and for each minute in an hour.

Still another sub-table gives entries for multiples of 56 years for cycles 1 to 20, and then for 40, . . . , 100, 200, 400, 600, 1000, . . . , 4000 cycles[!]. The first entry (0s 3;32°) may be computed by subtracting the value for year 1 (8s 3;5°) from the value for year 57 (8s 6;37°).

TV 11. Table for the motion of the lunar nodes (f. 109r)

This table lists the longitude of the lunar ascending node for each year in a period of 56 years. The entries are given in signs, degrees, and minutes. The value for year 57 is also given. Note that this period has no significance for the motion of the lunar node, and has simply been adopted on account of the previous table.

In the main table, the entry for year 1 is 3s 16;17°. As was the case in the previous table, this value corresponds to March 1, 1461. According to the Alfonsine Tables, the longitude of the ascending node at noon on the day before March 1, 1461 at Salamanca is 106;17,22°, in full agreement with the tabulated value. Note that this table and the previous one are both concerned with the lunar nodes and begin in March.

Other sub-tables show entries, also given in signs, degrees, and minutes, for each hour in a day, and for each minute in an hour. All other entries in the sub-tables for the motion of the lunar node rely on the basic Alfonsine parameter: −0;3,10,38,7,14°/d.

Still another sub-table gives entries for multiples of 56 years for cycles 1 to 20, and then for 40, . . . , 100, 200, 400, 600, 1000, . . . , 4000 cycles[!], once again. The first entry (0s 3;7°) may be obtained from the value for year 1 (3s 16;17°) and the value for year 57 (3s 13;10°) in the main table.

TV 12. Table for parallax (ff. 109v-110r)

The heading indicates that this table was computed for a latitude of 41;19°, which corresponds to the city of Salamanca, although its name is not given in the heading.

The presentation of this table is the same as in Ptolemy's *Handy Tables*.[3] It is composed of a set of seven sub-tables arranged for the moments when the Sun enters the signs of Cancer (12 June); Leo (14 July) and Gemini (12 May); Virgo (14 August) and Taurus (12 April); Libra (14 September) and Aries (12 March); Scorpio (14 October) and Pisces (9 February); Sagittarius (13 November) and Aquarius (10 January); and Capricorn (12 December).

In each monthly table the first column lists the local time for each integer hour from sunrise to sunset as well as the time for nonagesimal, the highest point of the ecliptic above the local horizon where the parallax in longitude vanishes. In this context, parallax means the adjusted parallax, i.e., the difference between the parallaxes of the Moon and the Sun. The second column has the heading *equatio temporis* and displays the component in longitude of adjusted parallax in hours and minutes. The same component is given in the third column, although here it is displayed in minutes of arc, with the heading *differencia longitudinis*. Finally, the entries in the fourth column correspond to the component in latitude of adjusted parallax, also in minutes of arc, with the heading *differencia latitudinis*. Note that the monthly tables are intended for use at the time of solar eclipses when it is reasonable to consider the adjusted parallax, and that they are not intended to be used for finding the lunar parallax at arbitrary times.

The times of sunset, counted from noon, are: 7;30h (June), 7;16h (May and July), 6;42h (April and August) and 6;0h (March and September), and their complements in 12h for the other months. They are coherent with the latitude for which the table was calculated, 41;19° (Salamanca).

The reported times of nonagesimal are: noon (Cancer), −0;24h (*sic*) (Leo), 0;58h (Virgo), 1;30h (Libra), 1;33h (Scorpio), 1;1h (Sagittarius), where we have adopted the convention that times before noon are negative and times after noon are positive. For the other zodiacal signs the same times with the opposite algebraic sign are displayed.

It is very unusual to present parallax in longitude in units of time, for most medieval tables display it in units of arc, as is the case for the parallax in latitude. The only other examples we have found are in Bonjorn's tables, where the longitudinal component is given in units of time, and

[3] Stahlman 1959, pp. 268–283; see also Chabás and Tihon 1993.

the latitudinal component as an arc (Chabás 1992, pp. 243–248); and in Peurbach's *Tabulae eclypsium* (1514), ff. e8v-f6r. In Table TV 12 both columns for the longitudinal component are related by a factor, 0;32,56°/h, equivalent to a mean lunar velocity of 13;10,35°/d.

TV 13. Table for the eclipsed fraction of the solar and lunar disks (ff. 110r, 111v)

The table on f. 110r gives the area digits (with a maximum of 12) of the eclipsed solar and lunar disks as a function of the linear digits (with a maximum of 12) of the eclipsed solar and lunar diameters, respectively. This table is already found in Ptolemy's *Almagest* (VI, 8) and in the *Handy Tables* (Stahlman 1959, p. 258), as well as in a number of earlier sets of tables used in medieval Spain.

On f. 111v there is a similar table, where the argument, the eclipsed diameter of the Sun and the Moon, is given in column 1 at intervals of 0;18 digits rather than at intervals of 1 digit (40 entries in all). Column 2 shows the corresponding fraction of the solar disk (*expansio solis*), and column 3, intended for the corresponding fraction of the lunar disk (*expansio lune*), is entirely blank.

TV 14. Table for lunar latitude (f. 110v)

This table is intended for the computation of eclipses. It gives the lunar latitude, called *latitudo lune precissa* in the heading, as a function of the argument of lunar latitude. The argument of latitude ranges from 11s 24° to 0s 17° and from 5s 13° to 6s 6° at intervals of 0;15°. Here the latitude is given to seconds. The entries are based on a maximum of 4;29°, rather than on the usual Ptolemaic value of 5;0° (cf. Table AP 18).

Very similar tables, with a maximum latitude of 4;30°, are found in al-Khwārizmī, Yaḥyā ibn Abī Manṣūr, and Levi ben Gerson.[4] But there are at least two Spanish zijes where we find a pair of tables for the lunar latitude, one with a maximum 5;0° and the other with 4;29°: *al-Muqtabis* of Ibn al-Kammād, and the Tables of Barcelona.[5]

TV 15. Table for solar eclipses (f. 110v)

This table gives the magnitude, in digits and minutes of a digit, and the half-duration, in minutes of time, of solar eclipses at mean distance as a

[4] For the table in al-Khwārizmī's zij, see Suter 1914, pp. 132–134, and Neugebauer 1962, pp. 95–98. For a reference to Yaḥyā's zij, see Kennedy 1956, p. 146; and for the table in Levi ben Gerson, see Goldstein 1974, pp. 132–133 and 212–217, col. VI.

[5] For Ibn al-Kammād's zij, see Chabás and Goldstein 1994, pp. 20–22; for the Tables of Barcelona, see Chabás 1996a, p. 504.

function of the difference between the latitude and the parallax in latitude of the Moon, $\beta - p_\beta$, given in minutes of arc from 0;0,0° to 0;22,0° at intervals of 0;0,30°, and from 0;22,0° to 0;28,15° at intervals of 0;0,15°, as well as for 0;28,23°.

For an explanation of its origin and the way it was computed, see Table AP 15 in the *Almanach Perpetuum*, but note that Table TV 15 has more than twice as many entries as AP 15.

TV 16. Table for lunar eclipses (f. 111r-v)

This tables gives the magnitude, in digits and minutes of a digit, the half-duration of the eclipse, and the half-duration of totality, in hours and minutes, of lunar eclipses at mean distance as a function of the argument of lunar latitude, given in minutes of arc from 5s 18;0° or 11s 18;0° to 6s 12;0° or 0s 12;0°, at intervals of 0;5°.

For an explanation of its origin and the way it was computed, see Table AP 16 in the *Almanach Perpetuum*, but note that Table TV 16 has six times as many entries as Table AP 16.

TV 17. Table for the correction of eclipses (f. 111v)

The heading for the table is *prima [tabula] continet equationem eclypsium ad epyciclum*. The argument is the lunar anomaly, and is given here at intervals of 2° from 0s 0° to 11s 30°. It is displayed in four columns, one for each quadrant. The entries for the correction are given to seconds under the heading *expansio solis*, and range from 0;6,0° at arguments 0° and 180°

Table TV 17A: Correction of eclipses

Lunar anomaly		*Expansio solis*	Lunar anomaly	
0s 0	0s 0	0; 6, 0	6s 0	6s 0
0s 2	11s 28	0; 5,59	6s 2	5s 28
0s 4	11s 26	0; 5,58	6s 4	5s 26
. . .				
1s 0	11s 0	0; 5, 0	7s 2*	5s 0
. . .				
2s 0	10s 0	0; 3, 0	8s 2*	4s 0
. . .				
2s 26	9s 4	0; 1, 1	8s 28*	3s 4
2s 28	9s 2	0; 1, 0	9s 0*	3s 2
3s 0	0s 0[a]	0; 0, 0	—	3s 0

a. Instead of 9s 0.

* The entry 6s 6 is missing in this column, and the remaining entries have thus been shifted upwards one place.

to 0;1,0° at arguments 90° and 270°. This table is partially reproduced in Table TV 17A.

This table derives ultimately from column 3 of Ptolemy's "table of corrections" in the *Handy Tables* (Stahlman 1959, p. 257), and raises several problems. They are discussed in detail in the comments to Table AP 13, where one of its columns seems to be derived from Table TV 17.

TV 18. Table for parallax at other latitudes (f. 112r)

This table, reproduced below as Table 18A, gives the corrections to be added to, or subtracted from, the component in longitude (col. 2, in minutes of arc, and col. 4, in minutes of time), and to (from) the component in latitude (col. 3, in minutes of arc) of the adjusted parallax for Salamanca as a function of the geographical latitude (cols. 1 and 5), when the observer is located in a range of latitudes roughly within 16° of latitude north or south of Salamanca ($\phi = 41;19°$). This complements Table TV 12, above, which gives the components of adjusted parallax for Salamanca. The headings for col. 2 *(equatio temporis)* and col. 4 (*diversitas latitudinis*) have been interchanged.

This is very precisely a tabular form of Zacut's "rough rule" (Heb. *derekh naqel*: an easy way): "There is an easy way to determine this very closely from the table I made for this *clima* [i.e., geographical latitude]. After you have found the conjunction, enter the table for the parallax as I instructed you above, and know the latitude for the place where you are . . . ." (*Ḥibbur*, at the end of chapter 6: MS B, ff. 21a–21b; cf. Cantera 1931, pp. 179–180). A summary of these instructions appears in chapter 8 of the Castilian version of the *Almanach Perpetuum* (Albuquerque 1986, p. 88). By "I made," Zacut probably refers to Table TV 12 rather than to the "rough rule" (Table TV 18), and thus he seems to claim authorship of a parallax table for Salamanca. The "rough rule" given in the *Ḥibbur* may be understood as follows: for each 1;15° of latitude above (below) the latitude of Salamanca—taken here roughly as 41;20°—add (subtract) 1 min of arc to (from) the component in latitude of parallax, and subtract (add) 2 min of time to the component of longitude expressed in units of time, or 1 min of arc if it is expressed in units of arc. Note that the table allows for a variation of 1 min in both components of parallax in units of arc when the variation of the latitude is 1° (not 1;15°, as in the text).

It is reasonable that someone who has gone through the very complex calculations needed to compute the entries in a table for parallax, thus understanding what is involved in it, and who owns a set of parallax tables for different latitudes should be able to find a simplifying algorithm, such as Table TV 18, to interpolate values between his and other latitudes. To illustrate how this could be achieved, and to have an idea of the accuracy of that rule, let us consider the set of tables for parallax in Polonius's *Tabulae*

Table TV 18A: Parallax at other latitudes

1	2	3	4	5
Lat.	Paral. lat. (′)	Paral. long. (′)	Paral. long. (min)	Lat.
20				
21				
22				
23				
24;20	17	17	31	58;20
25;20	16	16	29	57;20
26;20	15	15	27	56;20
27;20	14	14	26	55;20
28;20	13	13	24	54;20
29;20	12	12	22	53;20
30;20	11	11	20	52;20
31;20	10	10	18	51;20
32;20	9	9	16	50;20
33;20	8	8	15	49;20
34;20	7	7	13	48;20
35;20	6	6	11	47;20
36;20	5	5	9	46;20
37;20	4	4	7	45;20
38;20	3	3	6	44;20
39;20	2	2	4	43;20
40;20	1	1	2	42;20

Resolutae for Salamanca in Oxford, Bodleian Library, MS Can. Misc. 27, ff. 83r-87v. The entries for the parallax in longitude and latitude, in minutes of arc, in the first sub-table (Cancer) of each *clima*, and for 5h before noon are as follows:

Latitude	Paral. long.	Paral. lat.
24; 5° (*clima* II)	45	14
30;40° (*clima* III)	43	19
36;24° (*clima* IV)	42	24
39;54° (Toledo)	41	27
41;44° (*clima* V)	40	29
45;22° (*clima* VI)	35	29
48° (*clima* VII)	32	31

As can be seen, a variation of 24° in the geographical latitude corresponds to variations of just 13 min in parallax in longitude and 17 min in parallax in latitude, instead of 24 min according to Table TV 18, or 19 min according to the "rough rule."

TV 19. Table of geographical coordinates (f. 112r)

This is a list of 58 cities, for each of which we are given its longitude and latitude (both in degrees and minutes). The list is not ordered by increasing longitudes or latitudes. It begins with Toledo (L = 28;30° and ϕ = 39;50°), and includes Salamanca (L = 25;46° and ϕ = 41;19°) among a total of 28 Spanish localities, some of which are not usually included in this type of list, such as Castronuño (in the province of Valladolid) and Villalpando (in the province of Zamora). The Middle East is represented by 12 place-names; Italy by 7 place-names; France by 4 place-names; the Maghrib and Germany each by 2 place-names; and Portugal, England, and Austria each by 1 place-name. This table was transcribed by Laguarda (1990, p. 110), and he attributed it to Diego de Torres, professor at the University of Salamanca in the 1480s. Curiously, Laguarda identifies Diego de Torres with Abraham Zacut (Laguarda 1990, p. 88), but this bizarre conjecture has no foundation.

It is worth noting that Oxford, Bodleian Library, MS Can. Misc. 27 (f. 118v) has an analogous list of geographical coordinates, with 69 longitudes but no latitudes at all (except for Salamanca), and without the names of the first 50 places (except for Salamanca), making the table rather useless (Chabás 1998). But the sequence of longitudes is almost the same as in Madrid, MS 3385, including those for Castronuño and Villalpando! This fact does not imply that one manuscript was copied from the other, but it certainly indicates that there is a very close relationship between them.

TV 20. Table to convert arc units into time units (f. 112r)

For each integer degree from 1° to 60°, the table gives the corresponding time, in hours and minutes, where 1° corresponds to 0;4h.

TV 21. Table for sexagesimal multiplication (ff. 112v-113r)

This is a double argument table for interpolation purposes; its heading is *Tabula minutorum proporcionalium*. The horizontal argument ranges from 60 to 0, and the vertical argument from 1 to 30. The entries were computed by multiplying the corresponding arguments for the rows and columns. The table is not complete, and a quarter of it (towards the right lower corner of f. 113r) has been replaced by another multiplication table with horizontal argument ranging from 30 to 0 and vertical argument from 60 to 30.

2.4 The "Tables in Castilian"

Madrid, Biblioteca Nacional, MS 3385, also contains an extensive set of tables all of whose headings are in Castilian (ff. 139r-153r). It is one of the earliest surviving sets of astronomical tables written in that language. All the tables concern the two luminaries; there are no tables for the planets. The year begins in January in all cases. Another feature of these tables (hereafter: "Tables in Castilian") is the use of Roman numerals for all entries and signs of 30°. In the comments below, the tables will be preceded by the *siglum* "TC."

The first six tables are treated here as a group: they all give mean motions, except for Table TC 4, which lists the solar and lunar corrections at syzygies. The underlying parameters and the dates of the radices are presented after the comments to Table TC 6.

TC 1. Table for the mean motion in elongation (ff. 139r-140r)

This table gives the elongation between the Sun and the Moon (*lo que se aluenga la luna del sol*) as a function of time. It has 7 sub-tables for collected years (*años cogidos o ayuntados*, corresponding to the Latin *anni collecti*), for 20 expanded years (*años derramados*, sometimes called *esparsidos*, corresponding to the Latin *anni expansi*), for each month of the Julian year (both for a leap year and for a common year), and for each day, hour, and minute. The entries are given in signs, degrees, and minutes.

The radix for the collected years is 6s 24;31°. There follow entries for 40, 60, . . . , 100, 200, . . . , 1,000, 2,000, . . . , 7,000 years. In the sub-table for expanded years, the letter "b" (for bissextile) is added next to each year that is a multiple of 4. Some of the entries in this table are: 12;11° (1 day), 4s 9;37° (1 Julian year of 365 days), and 4s 13;25° (20 years).

TC 2. Table for the mean motion in solar anomaly (ff. 140v-141v)

This table displays the argument of solar anomaly (*argumento del sol*) as a function of time. As was the case for the elongation, it has 7 sub-tables, but for the collected years there are values for years 1460, 1480, 1500, . . . , 1780. The entry for 1460 is 6s 18;16°. The rest of the sub-tables follow the same pattern as those in Table TC 1. Some of the entries in this table are: 0;59° (1 day), 11s 29;45° (1 Julian year), and 11s 29;57° (20 years).

The entry for 1460 is the same as that found in the corresponding table (TV 6) in the *Tabule Verificate* for Salamanca preserved in the same manuscript (f. 105v). This means that Table TC 2 was also computed for Salamanca, and for January 1, 1461 as epoch. Note also that the collected years used here refer to completed years.

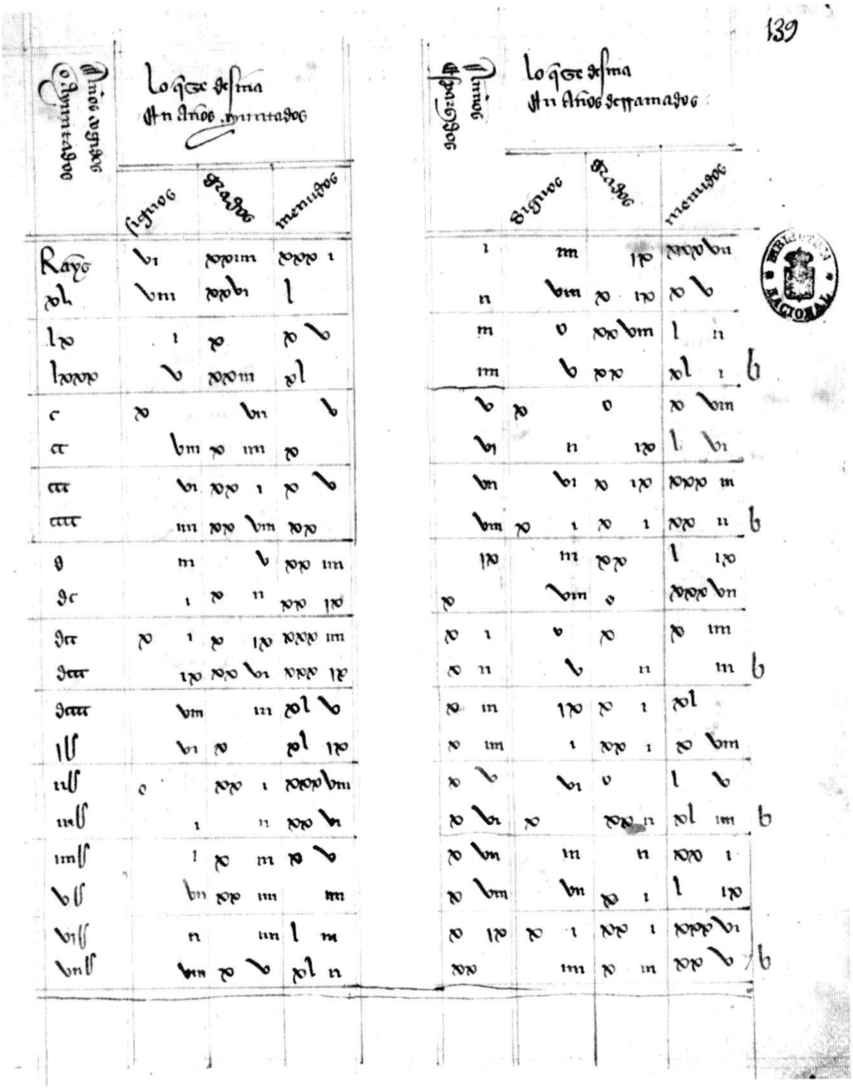

Figure 4. Table TC 1: Table for the mean motion in elongation (Madrid, Biblioteca Nacional, MS 3385, f. 139r).

TC 3. Table for the mean motion in lunar anomaly (ff. 142r-143r)

This table displays the argument of lunar anomaly (*argumento dela luna*) as a function of time. It is also presented as 7 sub-tables, with exactly the same format as Table TC 1. The radix for the collected years is 6s 19;6°. Some of the entries in this table are: 13;4° (1 day), 2s 28;43° (1 Julian year), and 1s 9;43° (20 years).

TC 4. Table for the solar and lunar corrections at syzygies (ff. 143v-144v)

The entries of this table are presented in 3 columns. Column 1 lists the argument for each integer degree from 1° to 180°, taken here as the solar or lunar anomaly. Columns 2 and 3 display respectively the solar and lunar corrections (*verificaçion*) in hours and minutes. Thus, they do not correspond to the usual solar and lunar equations, but rather represent the contribution in time of each luminary to be added or subtracted to the time of mean syzygy to obtain that of true syzygy.

The entries in col. 2 are the sum of those in cols. 2 and 4 of the corresponding tables in the *Tabule Verificate* (Table TV 8, f. 107r-v), ultimately derived from a table compiled by Nicholaus de Heybech of Erfurt (ca. 1400) to compute true syzygies (see Chabás and Goldstein 1992, pp. 265–289). The entries in col. 3 are the same as those in col. 2 of the table for the lunar correction in the *Tabule Verificate* (Table TV 7, f. 106r-v), and correspond to col. IV in Heybech's table. Note, however, that Table TC 4 cannot be derived directly from Tables TV 7 and TV 8; rather, it depends on the original table of Heybech.

TC 5. Table for the mean motion in solar longitude (ff. 145r-146r)

This table displays the mean motion of the Sun (*movimiento del sol*). It is also presented as 7 sub-tables, with exactly the same format as Table TC 1. The radix for the collected years is 9s 8;21°. Some of the entries in this table are: 0;59° (1 day), 11s 29;46° (1 Julian year), and 0s 0; 9° (20 years).

TC 6. Table for the mean motion of the ascending node (ff. 146v-147v)

This table gives the mean motion of the ascending node (*cabeça del dragon*), and again it has the same presentation as Table TC 1, except that the sub-table for minutes of an hour is lacking all headings and entries, and the frame is blank. The radix for the collected years is 3s 1;56°. Some of the entries in this table are: 0;3° (1 day), 0s 19;20° (1 Julian year), and 0s 26;50° (20 years).

The radices for the mean motions given in Tables TC 1 (elongation: 6s 24;31°), TC 3 (lunar anomaly: 6s 19;6°), TC 5 (solar longitude: 9s 8;21°), and TC 6 (ascending node: 3s 1;56°) should refer to a single date, and it turns out that they agree, but for scribal errors, with those for the Incarnation appearing in the Alfonsine Tables (see, for example, the *editio princeps*, 1483, c8r-v):

elongation	3,24;25,49,46,12°
lunar anomaly	3,19; 0,14,31,17°
solar longitude	4,38;21, 0,30,28°
ascending node	1,31;55,52,41°

Therefore, these tables depend on Alfonsine material for the radices, both for the Incarnation and for January 1, 1461. Moreover, all the mean motions parameters are also Alfonsine. Hence, the "Tables in Castilian" are related to the *Tabulae Resolutae* for Salamanca by Nicholaus Polonius, who introduced the Alfonsine corpus when he arrived there, no later than 1460. Now, the presence of some tables based on Heybech's table for syzygies that also depends on Alfonsine material indicates that the "Tables in Castilian" were written in the same intellectual milieu as the *Tabule Verificate* for Salamanca, a set that also contains Heybech's table.

TC 7. Solar equation (ff. 148r-148v)

This table displays the solar equation, again called *verificaçion*, as a function of solar anomaly. The entries are given in degrees and minutes from 1° to 180°, and reach a maximum of 2;10° for values of the argument between 90° and 97°. As expected, this maximum is an Alfonsine parameter (2;10,0°); moreover, all entries in this table seem to be rounded values of the entries in the standard Alfonsine table for the solar equation.

TC 8. Parallax (ff. 149r-151v)

There are 12 sub-tables, one for each zodiacal sign, beginning with Cancer. These tables are arranged in the same way as in Ptolemy's *Handy Tables* with only 3 columns. The first column displays the local time, expressed as an integer number of hours before or after midday, except for the first and last entries that give the times of sunrise and sunset (in hours and minutes). Columns 2 and 3 are headed *longura* and *anchura*, and give, respectively, the components of parallax in longitude (in hours and minutes) and in latitude (in degrees and minutes). In each table we are also given the time of nonagesimal, which is the highest point on the ecliptic above the local horizon, where the parallax in longitude vanishes. This information was probably added later to the manuscript, and the entries are written in Hindu-Arabic numerals, and not in Roman numerals as elsewhere in these tables: they are displayed in Tables TC 8A to 8F below the data corresponding to the nonagesimal within parentheses.

The times of sunset, counted from noon, are: 7;30h (Cancer), 7;15h (Leo and Gemini), 6;41h (Virgo and Taurus) and 6;0h (Libra and Aries), and their complements in 12h for the other signs, except for Sagittarius and Aquarius, which have 4;44h (instead of 4;45h). These times are coherent with the latitude of Salamanca (41;19°), but do not agree exactly with those in the *Tabulae Resolutae* or those in the *Tabule Verificate* (Table TV 12) also intended for Salamanca.

The times of nonagesimal are given as: noon (Cancer), 0;19h (Leo), 0;44h (Virgo), 1;17h (Libra), 1;40h (Scorpio), 1;24h (Sagittarius) after noon; and the same amounts of time before noon are given for the other zodiacal signs. These times differ significantly from those in the *Tabule Verificate* for Salamanca (Table TV 12).

Tables TC 8A through 8F reproduce the 12 sub-tables for parallax (ff. 149r-151v).

Table TC 8A: Parallax (Cancer and Leo)

Cancer time (h)	long. (h)	lat. (°)	Leo time (h)	long. (h)	lat. (°)
7;30	1; 0	—	7;15	1;20	—
7	1;10	—	7	1;22	—
6	1;12	6;31	6	1;25	5;10
5	1;13	6; 8	5	1;27	4;38
4	1;13	5;30	4	1;24	3;55
3	1; 7	4;36	3	1;17	3;27
2	0;50	3;28	2	0;55	2;40
1	0;23	3;10	1	0;36	2;30
noon	0; 0	2;52	noon	0;12	3; 4
1	0;23	3;10	(0;19	0; 0	3;19)
2	0;50	3;28	1	0;13	4; 2
3	1; 7	4;36	2	0;34	5;10
4	1;13	5;30	3	0;49	6; 8
5	1;13	6; 8	4	0;57	7; 8
6	1;12	6;31	5	1; 0	7;40
7	1;10	—	6	0;54	7; 3
7;30	1; 0	—	7	0;47	—
			7;15	0;47	—

Table TC 8B: Parallax (Virgo and Libra)

Virgo time (h)	long. (h)	lat. (°)	Libra time (h)	long. (h)	lat. (°)
6;41	1;20	—			
6	1;27	—	6;0	1;30	—
5	1;30	3; 0	5	1;34	—
4	1;28	3; 0	4	1;28	2;55
3	1;15	2;52	3	1;16	2;55
2	1;12	2;52	2	1;12	3;27
1	0;48	3;15	1	0;50	4;24
noon	0;27	4;13	noon	0;28	5;33
(0;44	0; 0	4; 0)	1	0;12	6;54
1	0;16	5;22	(1;17	0; 0	7; 0)
2	0;19	6; 8	2	0;13	7;50
3	0;36	7;40	3	0;26	8;30
4	0;47	8;25	4	0;37	9; 0
5	0;49	9; 0	5	0;41	9;12
6	0;47	9;10	6;0	0;37	9; 9
6;41	0;45	—			

Table TC 8C: Parallax (Scorpio and Sagittarius)

Scorpio time (h)	long. (h)	lat. (°)	Sagittarius time (h)	long. (h)	lat. (°)
5;19	1;30	—			
5	1;35	—	4;44	1;19	—
4	1;29	3;30	4	1;19	5;30
3	1;17	3; 0	3	1;18	5;45
2	1;13	4;35	2	1; 1	6;31
1	0;51	5; 9	1	0;38	7;40
noon	0;28	7; 0	noon	0;20	8;25
1	0;13	7;50	1	0;16	9;12
(1;40	0; 0	9; 0)	(1;24	0; 0	10; 0)
2	0;13	8;51	2	0;20	9;22
3	0;26	9;12	3	0;35	8;50
4	0;38	9;32	4	0;49	8;36
5	0;42	9;12	4;44	0;50	—
5;19	0;37	—			

Table TC 8D: Parallax (Capricorn and Aquarius)

Capricorn time (h)	long. (h)	lat. (°)	Aquarius time (h)	long. (h)	lat. (°)
			4;44	0;50	—
4;30	1;15	—	4	0;49	8;36
4	1;12	5;40	3	0;35	8;50
3	1; 4	7;20	2	0;20	9;22
2	0;45	8;26	(1;24	0; 0	10; 0)
1	0;25	8;40	1	0;16	9;12
noon	0; 0	9;24	noon	0;20	8;25
1	0;25	7;50	1	0;38	7;40
2	0;45	8;51	2	1; 1	6;31
3	1; 4	9;12	3	1;18	5;45
4	1;12	9;32	4	1;19	5;30
4;30	1;15	—	4;44	1;19	—

Table TC 8E: Parallax (Pisces and Aries)

Pisces time (h)	long. (h)	lat. (°)	Aries time (h)	long. (h)	lat. (°)
5;19	0;37	—	6;0	0;37	9;20
5	0;42	9;12	5	0;41	9;12
4	0;38	9;32	4	0;37	9; 0
3	0;26	9;12	3	0;26	8;30
2	0;13	8;51	2	0;13	7;50
(1;40	0; 0	9; 0)	(1;17	0; 0	7; 0)
1	0;13	7;50	1	0;12	6;54
noon	0;28	7; 0	noon	0;28	5;33
1	0;51	5; 9	1	0;50	4;24
2	1;13	4;35	2	1;12	3;27
3	1;17	3; 0	3	1;16	2;55
4	1;29	3;30	4	1;28	2;55
5	1;35	—	5	1;34	—
5;19	1;30	—	6;0	1;30	—

Table TC 8F: Parallax (Taurus and Gemini)

Taurus time (h)	long. (h)	lat. (°)	**Gemini** time (h)	long. (h)	lat. (°)
			7;15	0;47	—
6;41	0;45	—	7	0;47	—
6	0;47	9;10	6	0;54	7; 3
5	0;49	9; 0	5	1; 0	7;40
4	0;47	8;25	4	1; 8	7; 8
3	0;37	7;40	3	0;49	6; 8
2	0;19	6; 8	2	0;34	5;10
1	0;16	5;22	1	0;12	4; 2
(0;44	0; 0	4; 0)	(0;19	0; 0	3;19)
noon	0;28	4;13	noon	0;12	3; 4
1	0;46	3;15	1	0;36	2;40
2	1;12	2;52	2	0;55	3;27
3	1;15	2;52	3	1;17	3;55
4	1;28	3; 0	4	1;24	4;38
5	1;30	3; 0	5	1;27	5;10
6	1;27	—	6	1;25	—
6;41	1;20	—	7	1;27	—
			7;15	1;20	—

TC 9. Solar eclipse (f. 152r)

This table displays the magnitude (in digits and minutes of a digit), and the half-duration (in minutes of time) of solar eclipses at mean distance of the Moon, as a function of the argument of lunar latitude (*argumento del ancho*). Despite the occurrence of "signs" in the heading (see Table TC 9A, below), the argument of lunar latitude is given in degrees. The maxima listed for the argument are corroborated in one of the various notes on f. 153r, all written in Castilian and Latin in the same hand as the tables but using Hindu-Arabic, not Roman, numerals. A few lines below, there is another short note refering to solar eclipses: *Para que sea eclipsi del sol son menester estas condiciones. La primera que el argumento del ancho al tiempo dela conjuncion verdadera caya dentro delos terminos del eclipse del sol: 5s 11g usque 6s 6g 4min, 11s 23g 56 usque 0s 18g. La segunda condicion que la conjuncion sea de dia e non de noche . . .* (A solar eclipse must fulfill these conditions. First, for true conjunction the argument of lunar latitude must lie between the limits for solar eclipse: from 161° to 186;4°, and from 353;56° to 18°. Second, conjunction must happen during the day, not at night . . .)

This table seems unrelated to those of Levi ben Gerson, Bonjorn, or the corresponding table in the *Ḥibbur* (the same as in the *Tabule Verificate*,

Table TC 9A: Solar eclipse

argument of lunar latitude		magnitude	half-duration
0	30	11;50	56
1	29	9;55	55
2	28	7;53	53
3	27	5;56	48
4	26	3;57	40
5	25	2; 3	31
6	24	0; 6	2
6;4	23;56	0; 0	—

and the *Almanach Perpetuum*). The tabulated function in the column for the magnitude should be linear, but the line-by-line differences (1;55, 2;2, 1;57, 1;59, 1;54, and 1;57) are not constant, thus indicating that the entries were miscomputed or miscopied. One can only guess that these entries were computed from a table for solar eclipses probably using the argument of lunar latitude (not the latitude) as argument, and containing sub-tables for the Sun at apogee and perigee, as is the case, for instance, in the *Almagest* (VI, 8) and in al-Khwārizmī's zij.

TC 10. Lunar eclipse (f. 152r)

As in the previous table, for each value of the argument of lunar latitude (in degrees), this table displays the magnitude of an eclipse (in digits and minutes of a digit), and the half-duration of an eclipse (in hours and minutes). This is again a table for eclipses at mean distance of the Moon.

The magnitudes of an eclipse for arguments from 7° to 12° are taken from the table for lunar eclipses at mean lunar distance compiled by Jacob ben David Bonjorn (Chabás 1992, p. 251). The entry for argument 6° should be 12;2, and the entries for 0° to 5° are obviously erroneous. The same is true for the entries for the half-duration of an eclipse: from 6° to 12° the entries are identical with those in Bonjorn's table, and the rest are erroneous and absurd.

Note that Bonjorn's table for lunar eclipses is also found in various forms in the *Ḥibbur* (Table HG 20), the *Tabule Verificate* (Table TV 16), and the *Almanach Perpetumm* (Table AP 16).

TC 11. Entry of the Sun into the zodiacal signs (f. 152v)

This table has its heading in Latin, and the accompanying notes are also in Latin. The entries are written in Hindu-Arabic, not Roman, numerals. A marginal note explicitly states that the table is valid for the meridian of

Table TC 10A: Lunar eclipse

argument of lunar latitude		magnitude	half-duration
0	30	12; 0	2;46
1	29	12; 0	2;44
2	28	12; 0	2;40
3	27	12; 0	2;35
4	26	12; 0	2;24
5	25	12; 0	2;10
6	24	12; 0	1;35
7	23	10; 1	1;29
8	22	8; 0	1;22
9	21	5;58	1;12
10	20	3;58	1; 1
11	19	1;58	0;43
12	18	0; 0	0; 0

Salamanca, 1473, in equated days (*ad meridianum Salamantinum diebus equatis anno domini 1473*).

In this table, the entries for the days and the hours for March through December agree with those in Tables HG 5 and AP 6, valid for 1473. This is not the case for the entries for the minutes and the seconds (except for March), which seem to have been corrected, possibly by another hand.

As mentioned above, f. 153r contains a set of notes written in Castilian

Table TC 11A: Entry of the Sun into the zodiacal signs

		d	h	m	s
Jan	Aqu	9	22	10	0
Feb	Psc	8	12	49	36
Mar	Ari	10	16	0	12
Apr	Tau	10	9	31	18
May	Gem	11	14	50	32
Jun	Cnc	12	3	25	18
Jul	Leo	13	16	18	16
Aug	Vir	13	23	35	24
Sep	Lib	13	16	36	48
Oct	Sco	13	20	38	12
Nov	Sgr	12	11	58	18
Dec	Cap	12	19	58	59

and Latin in the same hand as that of the tables. There are a few references to the *Tabulae regis Alfonsi* and the only locality mentioned is Salamanca.

In sum, we can say that these "Tables in Castilian" were intended for Salamanca, and one of them has Jan. 1, 1461 as epoch. They depend on material from the Alfonsine corpus, both standard tables and others such as Nicholaus de Heybech's tables. But they also contain non-Alfonsine material, such as Bonjorn's tables, that are frequently found both in Hebrew and Latin in the astronomical literature in Spain during the fifteenth century. The "Tables in Castilian" bear many similarities to other tables compiled for Salamanca at about the same time: the *Tabulae Resolutae*, the *Tabule Verificate*, and Zacut's *Ḥibbur*, and help us to understand more fully astronomical activity in that Castilian city. Based on the evidence presented here, it is not possible to assign authorship to either the *Tabule Verificate* or the "Tables in Castilian," even though there are several very good candidates: Juan de Salaya, Diego Ortiz de Calçadilla, and Fernando de Fontiveros, all of whom held the chair of astronomy/astrology at the University of Salamanca in succession from 1464 to 1480, worked in the same astronomical milieu, and had access to the *Tabulae Resolutae* of Nicholaus Polonius, the first incumbent of the chair.

2.5 Other material related to Salamanca

TOLEDO, BIBLIOTECA DE LA CATEDRAL, MS 98–27, is a fifteenth century codex of 130 folios, containing various scientific treatises, such as the *Theorica Planetarum*, beginning with *Circulus ecentricus egresse cuspidis*; the *Practica astrolabii,* beginning with *Primum horum est armilla*; the *Aritmetica speculativa* by Thomas of Bradwardine; and Book I of Euclid's *Elements*. (For a description of this manuscript, see Millás 1942, pp. 218–221.) On the spine of this volume we find "Antonius Lebrixa," the humanist and grammarian Elio Antonio de Nebrija (1444–1522), who was a student at Salamanca, spent ten years in Bologna, then some years at the court of the master of the Order of Alcántara, Juan de Zúñiga (at a time when Abraham Zacut was there), and later became a professor at the University of Salamanca. This suggests that this volume belonged to Nebrija's library, for neither his name nor any of his works is mentioned anywhere else in the manuscript. In a table of geographical coordinates for 54 cities (f. 120v), we note that the last entry is "Lebrixa," the hometown of Antonio de Nebrija. On f. 66r there is a small text on prognostications for 1454, and on f. 75r we find two astrological charts or horoscopes, labeled: *ad meridianum civitatis salamantine*, for August 24, 1458 at 3;45h p.m., and [August] 25, 1458 at 4;53h p.m. The first date corresponds to an opposition of the two luminaries, but the second does not seem related to any special configuration of the planets. On f. 123r there is a heading:

Canones super tabulas nunc noviter factas a quodam super civitate cuius longitudo est ab occidente vero 22;58° et ab equinoctiali latitudo 39;38°. Beaujouan (1969, p. 13) was quite right to identify this place with Lisbon (see the table of cities on f. 120v, mentioned above, where the coordinates of Lisbon are given), but we find no evidence there to support the claim that some tables were calculated at Salamanca around the year 1460 for the coordinates of Lisbon (Beaujouan 1967, p. 35). Moreover, what follows this heading is a disjointed set of notes on planets, with no tables at all (ff. 123r-124v).

On the other hand, on f. 130r there is a list of 12 quantities, under the heading *medii motus pro anno Christi completo 1460 ad finem*:

0s 10;43,35,31	auges
2s 14;17,42,35	access and recess
9s 19; 5, 2,58	Sun, Venus and Mercury
5s 2;59,42,28	Moon
7s 10;10,14,16	lunar anomaly
1s 13;20,54,16	argument of lunar latitude
8s 10;35,12, 7	ascending node
9s 25;43,30,11	Saturn (25°, erroneously for 26°)
7s 19;23,32, 0	Jupiter
4s 28;30, 6, 3	Mars
7s 5;38,17,56	center of Venus
0s 21; 3,35,29	center of Mercury

In the comments to Table TV 5, above, for the mean motion in lunar anomaly we found that when computing the lunar anomaly for the longitude of Salamanca, for January 1, 1461 as epoch, and using Polonius's *Tabulae Resolutae* for Salamanca, the result was 7s 10;10,14,36°, in very close agreement with the value listed on f. 130r. All other quantities correspond very precisely to computations made with the Alfonsine Tables for January 1, 1461, and the coordinates of Salamanca, thus providing additional evidence for the use of the Alfonsine Tables in Salamanca no later than 1461.

Although not bound together with the rest of the *Tabulae Resolutae*, Oxford, MS Can. Misc. 27 contains other tables related to Salamanca. One is for the entry of the Sun into each zodiacal sign (f. 118r). Its heading is *Tabula introituum solis in 12 signa et ascensiones ascendentis ad iddem (sic) tempus facta ad Salamanticam anno currente 1461* (see Table 2). Comparing both columns for time, it is readily seen that in computing this particular table, the author used the table for the equation of time in the

Table 2: Entry into the signs at Salamanca (1461)

Nomina signorum	*Nomina mensium*	*Tempus diebus non equatis* d h m 2 3	*Tempus diebus equatis*	*Ascensiones ascendentis*
Aqu	Jan	10 0 9 10 0	10 0 11 58	35 12 30
Psc	Feb	8 14 54 58 23	8 14 54 22	285 57 30
Ari	Mar	10 18 0 30 1	10 18 8 38	2 9 30
Tau	Apr	10 11 6 47 0	10 11 24 43	289 3 45
Gem	May	11 16 45 33 0	11 17 6 53	44 30 15
Cnc	Jun	12 5 28 38 0	12 5 45 6	266 16 30
Leo	Jul	13 18 24 50 42	13 18 36 19	131 17 45
Vir	Aug	14 0 36 13 46	14 0 50 42	254 47 30
Lib	Sep	13 18 25 44 33	13 18 49 44	192 26 0
Sco	Oct	13 22 17 42 48	13 22 49 7	280 9 45
Sgr	Nov	12 13 38 16 34	12 14 6 57	179 31 15
Cap	Dec	11 21 54 4 36	11 22 9 43	332 25 45
Sco 20°	Nov	1 17 59 37 0	1 18 30 49	234 13 15
Sco 21°	Nov	2 17 59 25 40	2 18 30 22	

Tabulae Resolutae (f. 79r), thus suggesting that this particular format of the Alfonsine Tables had reached Salamanca no later than 1461.

Note that in the case of Scorpio the table gives the times for the entry of the Sun into two specific degrees rather than into the sign.

2.6 Predecessors explicitly acknowledged by Zacut

ZACUT WAS VERY WELL INFORMED of the astronomical work by his Jewish predecessors, and mentions a great many of them, notably: Abraham Ibn ʿEzra (12th c.), Moses Maimonides (12th c.), Jacob ben Tibbon (13th c.), Isaac ben Sid (13th c.), Isaac Israeli (14th c.), Levi ben Gerson (14th c.), Jacob ben David Bonjorn (or Poel, 14th c.), Immanuel ben Jacob Bonfils (14th c.), and Isaac al-Ḥadib (14th c.). There are three other astronomers cited by Zacut for whom we can offer some new information, based on recent research: Judah ben Asher II (14th c.), Ḥayyim of Briviesca (14th c.), and Judah ben Verga (15th c.).

At the beginning of the fourteenth century, Asher ben Yeḥiel (d. 1328) became the chief rabbi in Toledo, and his descendants were important leaders of the Jewish community in Spain throughout that century. Among them was Judah ben Asher (d. 1391), Asher ben Yeḥiel's great grandson, often confused with the better known son of Asher ben Yeḥiel by the same name (d. 1349). The astronomer frequently cited by Zacut is undoubtedly the great grandson, and we will refer to him as Judah ben Asher II, or simply as Judah ben Asher. Thanks to Y. T. Langermann, the tables of Judah ben Asher II have been identified in Vatican, MS Heb. 384. On f. 284a the author's name is given only as "Judah" without further specification, but we can identify this author as Judah ben Asher II. Moreover, the title in the Vatican manuscript is given as *Ḥuqqot shamayim* (*Statutes of the heavens*: cf. Jer. 33:25), and this is the also the title of Judah ben Asher's treatise cited by Zacut in chapter 18 (Cantera 1931, p. 340). In his *Sefer Yuḥasin* (Alfred Freimann 1924, p. 225a), Zacut tells us that "at that time there was in Burgos a great scholar who, like his ancestors, knew the entire Talmud, the scholar R. Judah ben Asher, the great-grandson of the *Rosh* [R. Asher ben Yeḥiel], and he was killed in Toledo in 5159 [= 1391], and he wrote the book, *Ḥuqqot shamayim*."

The tables are uniquely preserved in this Vatican manuscript, and unfortunately it is in a poor state of preservation. In the canons to these tables the city of Burgos and the year 1364 are mentioned (ff. 285a, 291a), and on f. 292a we are told: "If you wish to find the radices of the planets for a city other than Burgos . . . ." The reference to Burgos is significant for us, because Judah ben Asher II is associated with this city in various documents (Alfred Freimann 1920, pp. 152 ff). Of particular interest is that Zacut has a double argument table for the unequal motion (i.e., the daily velocity) of Mercury (Table AP 42) that was probably copied from Judah ben Asher (although it is possible that they both depended on a common source). Judah ben Asher's Tables include such double argument tables for all five planets as well as the Moon, whereas Zacut only has a table of this kind for Mercury. Moreover, Zacut's table has entries in degrees and minutes, while the corresponding table for Mercury's daily velocity in Vatican, Heb. 384, gives values to seconds which, when rounded, yield the entries in Zacut's table (Goldstein 1998, p. 180).

There is not much known about this Ḥayyim of Briviesca (in the province of Burgos in Spain). In London, British Library, MS Or. 10725, ff. 12a-14b, there are notes on the canons of Levi ben Gerson's tables by Ḥayyim of Briviesca. On f. 12a there is a worked example for a time specified in four ways: 5149 A.M.; 1389 A.D.; 1427 Spanish Era (Era Caesar: epoch −37); and year 69 after 1320, the radix of the tables (i.e., Levi's circle 1: cf. Goldstein 1974, pp. 134, 218). Note that the introduction to Levi's tables appears in this MS on ff. 1b to 10b, and Levi's tables themselves on ff. 29a-35b. In chap. 9 of Zacut's *Ḥibbur,* Ḥayyim is mentioned (MS B, f. 26b:−2; MS L, f. 203r; MS S, f. 31v; see Cantera 1931, pp. 191, 292 n. 661).[6] Here Ḥayyim is associated with Judah ben Asher of Burgos (d. 1391), and it would seem that they were contemporaries who

[6] For an explanation of manuscript sigla, see p. 53.

lived near one another. There is also a passage in a supplement to the *Ḥibbur* that Cantera did not report (MS B, f. 27b; MS W, f. 12b), where we are told that: "The scholar, Abraham Zacut said: I did this at first relying on the table that R. Ḥayyim of Briviesca constructed, 'The table for finding the ascensions for all places' but later, upon reflection, I decided that it was inappropriate to rely on this [table] since he constructed it with a precision of minutes only, and from it a considerable error may result. . . . Therefore, I decided (MS B: *lit.* returned; MS W: *lit.* came) to construct this table to a precision of seconds. . . ."

Ḥayyim also wrote a supercommentary on Ibn ʿEzra's commentary on the *Pentateuch*, called *Eṣ ḥayyim* (*The Tree of Life*), preserved in Leeuwarden (The Netherlands), MS Heb. 5. In it Ḥayyim displays his knowledge of astronomy and his interest in mysticism. Among the many marginal notes, there are often references to *mori R. Levi* (my teacher, Rabbi Levi); Steinschneider (1964, p. 177), and others following him, assumed that this Levi is Levi ben Gerson (d. 1344). But a recent examination of the manuscript indicates that this is not the case, for elsewhere there is a reference to Levi ben Gerson (with his full name), followed by "may his memory be for a blessing" (indicating that he was deceased at the time this was written). Moreover, there is a verifiable quotation from Levi ben Gerson's commentary on the *Pentateuch*, whereas the citations of "my teacher, R. Levi" are not consistent with Levi ben Gerson's commentary (privately communicated by Y. T. Langermann).

Judah ben Verga is associated with Lisbon, and was active from about 1455 to 1480 (Langermann 1999, pp. 18–25, 34). Until recently, Zacut's references to Ben Verga's astronomical tables were not supported by any extant text by him. But Y. T. Langermann has now identified a set of tables with canons that were compiled by Ben Verga (canons in: St. Petersburg, MS Heb. C-076, ff. 57a-65a; tables in: Paris, Bibliothèque Nationale de France, MS Heb. 1085, ff. 86b-98a, and Oxford, Bodleian Library, MS Poc. 368, ff. 222b-236b [cf. Neubauer 1886, MS Nb. 2044]). In the canons, the author says that he was in Lisbon (St. Petersburg, f. 57a), and in one copy of the tables, Lisbon is mentioned in the heading of a table (Oxford, f. 232a); but Ben Verga's name does not appear in either the canons or the tables. Moreover, in the canons we are also told that, in Lisbon, "I found that the autumnal equinox (*tequfat tishri*) took place after 13 completed days of September in the year 1456 of the Christians" (St. Petersburg, f. 64a); elsewhere, in Ben Verga's work called *The Instrument of the horizon* (*Keli ha-ofeq*), the very same observation is reported in the first person (Paris, Bibliothèque Nationale de France, MS Heb. 1031, f. 162a; see also London, British Library, MS Heb. Add. 27,106, f. 30a, cited in Margoliouth 1915, p. 342). Hence we are confident that Ben Verga is the author of these canons. In the tables (Oxford, f. 227a; Paris, f. 91a), the maximum solar equation is 1;53° which is the same as in the zij of Ibn al-Kammād, rounded to minutes (Chabás and Goldstein 1994, p. 5), and it is quite different from the value, 2;10°, in the Alfonsine Tables. Further, in the lunar correction table, where the argument is the anomaly (rounded to minutes) given at intervals of 1 day, the entries in the column labeled, "0 days," i.e., 0 days after syzygy, or 0° of double elongation, agree with values computed from the zij of al-Battānī. For example, for 7 days the argument of lunar anomaly is 3s 1;27° and the corresponding entry for 0° of double elongation is 5;1° (Oxford, f. 230b; Paris, f. 94b), whereas al-Battānī's zij yields 5;0,34° (Nallino 1899–1907, 2:78 ff). Again, this value is not compatible with the Alfonsine Tables where the maximum lunar equation for 0° of double elongation is 4;56°. So it seems that, unlike Zacut, Ben Verga did not use the Alfonsine Tables.

There are not many Greek and Muslim scholars mentioned in the *Ḥibbur*: Hipparchus (2nd c. B.C.), Menelaus (*ca.* 100 A.D.), Ptolemy (2nd c.), al-Farghānī (9th c.), al-Battānī (late 9th c.), al-Ṣūfī (10th c.), Azarquiel (11th c.), ʿAlī ibn Abī l-Rijāl (11th c.), and Averroes (12th c.). This short list suggests that Zacut's knowledge of Greek astronomy was limited to Ptolemy's *Almagest* (for both Hipparchus and Menelaus are mentioned in it), and that his access to Arabic sources was not extensive.

There are even fewer Christian predecessors mentioned in the *Ḥibbur*, namely, King Alfonso (d. 1284), and John of Lignères (ca. 1330) who is only mentioned in MS B. It is noteworthy that Zacut refers to none of the astronomers in Salamanca whose works, we argue, he consulted.

In the Canons to the *Almanach Perpetuum*, hardly any predecessors of Zacut are mentioned: 1 Greek (Ptolemy), 3 Muslims (al-Farghānī, al-Battānī, and Averroes), 1 Jew (Jacob ben David Bonjorn), and 2 Christians (Alfonso and John of Lignères). Clearly, a comparison of these lists indicates the distance that separates the *Ḥibbur* and the *Almanach Perpetuum*, and supports our claim that these are quite distinct works.

3. The *Ḥibbur*

3.1 The Tables

For the tables of the *Ḥibbur*, we have consulted 3 Hebrew MSS: L (Lyon, MS Heb. 14), the best witness for these tables; Mu (Munich, MS Heb. 109), W (Warsaw, ZIH, MS Heb. 245 [formerly Vienna, MS Heb. 301]); and 3 Latin MSS: Ac (Madrid, Academia de la Historia, MS Heb. 14); Ma (Madrid, Biblioteca Nacional, MS 3385); and Se (Segovia, Cathedral, MS 110). MS W has been described at length in Schwarz *et al.* 1973, pp. 40–52. On occasion we have consulted MS B (Oxford, Bodleian, Opp. Add. 8° 42), dated 1489, that contains Zacut's canons, but not the tables.[1] Similarly, we have consulted MS S (Salamanca, sign. 2–163), as transcribed by Cantera (1931, pp. 151–236), a Castilian translation of the canons completed in 1481 by Juan de Salaya with the help of Zacut. The manuscript at the Academia de la Historia, Madrid, is catalogued as a Hebrew manuscript although it contains no Hebrew at all: the headings are in Latin and the notes are in Castilian; a short notice of it was given by Cantera (1959, pp. 34–35). MS Ma has another set of tables that complements the tables in the *Ḥibbur,* entitled *Tabule Verificate*: it has March 1, 1461 as epoch, and should be understood as pre-*Ḥibbur* material (see chapter 2).

The most extensive set of these tables in Latin is preserved in MS Ac, and it is particularly noteworthy that MSS Ac and L are the only witnesses for Zacut's unusual base-30 tables (see chap. 3.2.1). It is difficult to establish the name of the Latin translator of the tables of the *Ḥibbur* because the manuscripts are silent on this matter. But it is tempting to consider that Zacut himself had some involvement, as he had in the translation of the canons of the *Ḥibbur* into Castilian by Juan de Salaya. In chapter 5.4 we will discuss some additional Latin material related to the *Ḥibbur* that indicates ongoing Christian interest in Zacut's work.

The tables in the *Ḥibbur* have been numbered here and each of them is preceded by the *siglum* "HG." References are given to the tables in the *Almanach Perpetuum*, to each of which we have given a number preceded by the *siglum* "AP."

[1] For a list of Hebrew MSS of the astronomical works of Zacut, see Goldstein 1981, pp. 246 ff. The manuscript listed there as N–19 (New York, JTSA, MS Heb. 296) can now be identified, for the most part, as a Hebrew version of the Almanac of 1307 (Chabás 1996b), with one table for the daily positions of the Sun in a 4-year cycle taken from Zacut's tables (Table HG 1, below). The folios in MS L have been assigned numbers as if it were a Latin manuscript; hence, they are in reverse order with respect to the Hebrew text.

HG	L	Mu	W	Ac	Ma	Se	AP
1.	185v-182r	48a-51b	21b-23a	1r-4v	1r-4v	15v-19r	2
2.	182r	52a	23b	5r	5r	19v	4
3.	181v	52a	—	5r	5r	19v	3
4.	181r	52b	23b	43r	41r	—	5
5.	180v-177r	53a-56b	19a-21a	6r-9v	5v-9r	—	6
6.	176v	57a	—	5v	—	—	—
7.	176r	57b	—	5r	5r	—	—
8.	176r	57b	24a	—	—	—	—
9.	175v-144v	60a-64b 66a-91b	24b-55a	10r-40v	9v-40r	21v-44v	7
10.	144r	58a-b	56b-58a	41r-v	104v-105r	—	9
11.	144r	58b	—	41v	—	—	10
12.	143v-r	184a-b	58b-59a	42r-v	40v, 109r	—	11
13.	142v	185a	24a	43r	41r	—	18
14.	142v-141r	185a-186b	56a-57a	43v, 45r	41v-42r	—	17
15.	140v-r	183a-b	59b	45v-46r	108r-109r	—	12
16.	139v-r	186b-187a	60a-b	46v	109v-110r	—	14
17.	139r	—	—	46v	111v	—	13
18.	138v	188a	61a	47r	110v	—	18
19.	138v	188a	61a	47r	110v	—	15
20.	138r	188b	61b	47v	111r-v	—	16
21.	138r	188b	61b	47v	—	—	—
22.	138r	188b	61b	47v	110r, 111v	—	15, 16
23.	137v-135r	92a-94b	16a-18b	48r-50v	63v-65r	—	1
24.	134v	164a	15b	44r	—	—	—
25.	134v	164a	15b	44r	—	—	—
26.	134r	164b	14b	44v, 53r	—	—	19
27.	133v	165a	15a	53v-54r	—	—	—
28.	133r	165b	—	54v	—	—	49
29.	132v	166a	—	—	—	—	—
30.	132r	211a	—	—	—	6v-7r	47
31.	132r	183a	—	—	—	7v-8r	47
32.	131v-115r	167a-182b	—	—	—	—	—
33.	114v-r	—	62a-b	—	101v	—	45
34.	113v-112r 102v	95a-97a	63a-65b	56r-60v	43r-44r	67v-68v	20
35.	110v-108v	97b-99b	66a-68a	120v-124v	67v-70v	—	21
36.	108r-104v	100a-103b	68b-71b	135r-139r	78r-81r	—	22

continued

HG	L	Mu	W	Ac	Ma	Se	AP
37.	104r-103r	104a-105a	72a-b 73a-75b	152v-153r 154r-v	91v-93r	—	23
38.	102r-97r	106a-110b	76a-82b	61r-70r	44v-48r	69r-79r	24
39.	94r	218a-b	—	—	84v	—	—
40.	93v	215a, 159a	—	109r	66r	—	—
41.	92v-89v	110b-114a	83a-86a	125r-131r	70v-75r	—	25
42.	89r-84r	114a-119a	86b-89b	139v-145v	81v-86r	—	26
43.	83v-82v	119b-120b	90a-b	153, 155r-156r	93v-94r	—	27
44.	82r-74v	121a-138a	92a-99b	71r-79r	49r-56r	63v-67r	28
45.	74r-72r	129a-130b	100a-101a	131v-133v	75v-77r	—	30
46.	72r-70r	131a-133a	101b-102b	146r-148r	86v-88r	—	31
47.	69v-68v	133b-134b	103a-b	156v-158r	94v-95r	—	32
48.	2r	—	104a	70v	—	—	29
49.	68r-64v	135a-138b	105a-108b	80r-85v	56v-60r	59r-63r	33
50.	64r	139b	109a	86r	60r	60r	34
51.	64r	139a	109a	134r	77v	—	35
52.	63v	139b-140a	109b	148v-149r	88v	—	36
53.	63r-60v	140b-143a	110a-111b	158v-161r	95v-97r	—	37
54.	60r-44v	143b-148a 148b	113a-126a	86v-105v	60v-63r	53r-58v	38
55.	44v	148b	—	—	—	—	39
56.	44r	148b	—	134r-v	77v	—	40
57.	44r-41v	148b-151a	—	149r-152r	89r-91r	—	41
58.	41r-38v	151b-154a	—	106r-108v	—	—	42
59.	38r-33v	154b-159a	—	161v	97v-101r	—	43
60.	—	159a	—	109r	—	—	—
61.	33r	159b	—	109v	42v	—	—
62.	32v-31v	160a-161a 211b-214b	—	110r-111r	—	—	44
63.	31r-29v	161b-163a	—	111v-113r	—	—	—
64.	29r-27v	189a-190b	—	113v-114v	—	—	—
65.	27r-25v	191a-192b	—	115v-116r	—	—	—

HG 1. "Table for the true position of the Sun: Epoch 1473"

This table for the daily solar positions displays the true longitudes of the Sun, given to seconds, calculated for the meridian of Salamanca for noon of each day of a four-year cycle beginning on January 1, 1473 (Latin MSS) or March 1, 1473 (Hebrew MSS). It is essentially the same as Table AP 2, although there are two slightly different versions of Table HG 1. One version is represented by MSS Mu and W (in Hebrew), and the other by MS L (in Hebrew) and MSS Ac, Ma, and Se (in Latin) . Both versions agree for all entries in the first year, but from March 1, 1474 to the end of the table there is a systematic difference in the seconds. These differences are sinusoidal, reaching a maximum when the solar longitude is at Can 0°, and a minimum at Cap 0°. We cannot explain this peculiarity in the table, but we think it unlikely that a proper astronomical explanation can be found for it.

HG 2. "Table for the solar correction for other revolutions"

In MS L there is a duplicate of this table on f. 181v. It is identical with Table AP 4 (solar correction), which gives the corrections to be added to the true position of the Sun after successive four-year cycles.

HG 3. "Table for the declination of the Sun and the 7 [*sic*] planets from the ecliptic according to al-Zarqāllu [MS L: Azarqel; MS Mu: al-Zarqel; MS W: the heading is on f. 57a — ruled for a table but without entries]"

This table is identical with Table AP 3 which gives the solar declination to minutes as a function of the solar longitude.

HG 4. "Table for the days and their nights in minutes of an hour"

This table is identical with Table AP 5 (equation of time), and begins in March.

This table is not found in MS Se which has a different table for the equation of time (ff. 46r-46v) where the entry for Cap 1° is 3;46°, thus differing from that in al-Battānī's zij, the Toledan Tables, and the *editio princeps* of the Alfonsine Tables (1483), where the entry for Cap 1° is 3;41°.

HG 5. "Table for finding the day when the Sun enters the beginning of each sign, and from it you may find the *tequfot*, and it is corrected for the equation of time, and the epoch is 1473"

This table is identical with Table AP 6 (entry of the Sun into each zodiacal sign). It is not found in MS Se, which in turn has four tables for the entry of the Sun into Aries, Cancer, Libra, and Capricorn, respectively, for 50 years, written in cistercian numerals, beginning in 1475 (See Chabás and Goldstein 1998). These four tables also give the positions of the 5 planets and the Moon, and the ascendants at the time of the entry of the Sun into the four zodiacal signs.

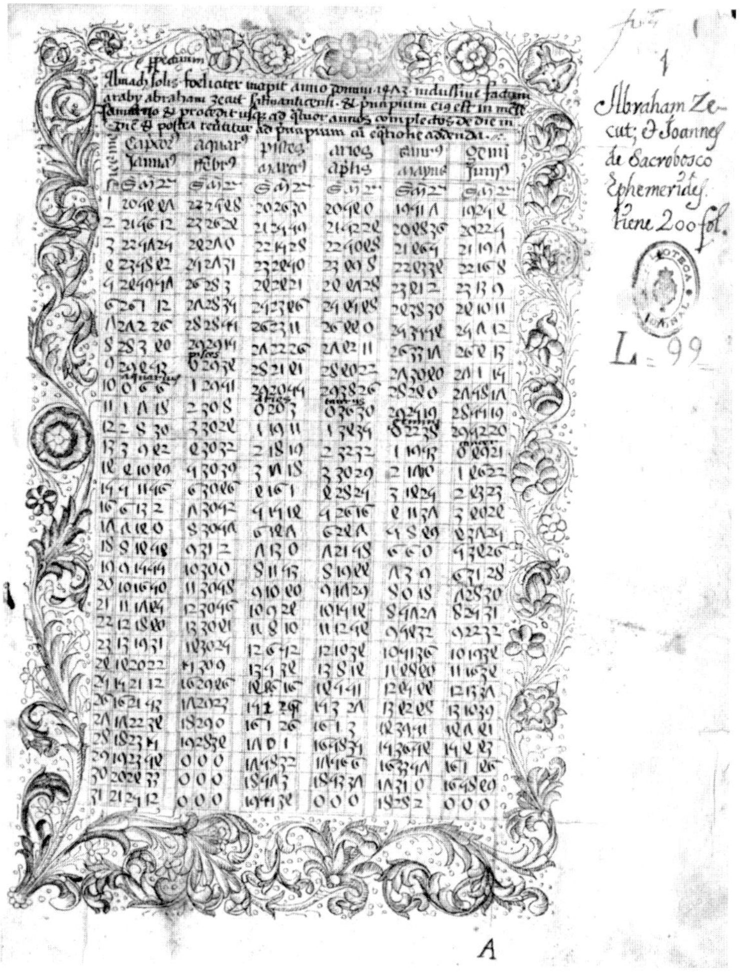

Figure 5. Table HG 1: Table for the true position of the Sun: Epoch 1473, first page only (Madrid, Biblioteca Nacional, MS 3385, f. 1r).

Table HG 6A: The Day at the Beginning of the Month

Year of the cycle	*Tishri* . . . . . . *Elul*
1	21* Sep . . . . . . 11 Aug
.	
.	
19	3 Sep . . . . . . 23 Aug

* 22 in MS Ac

HG 6. "Table for finding the day of the Christian month corresponding to the beginning of the Jewish month"

This table corresponds closely to Isaac Israeli's Table XL (Goldberg and Rosenkranz (eds.) 1846–48), where year 1 begins in September 1294 (the first year of 19-year cycle 267, or 5055 A.M.), but it is undoubtedly for another cycle. The table can be deduced from Levi ben Gerson's Table 10 (Goldstein 1974, p. 168), for it is its reverse. It is not found in the *Almanach Perpetuum*. In MS Ac a marginal note adds: "The radix of this table is the beginning of 5226 A.M., that is, 1466 A.D." In fact, 5226 A.M. corresponds to 1465–66 A.D., and is the beginning of 19-year cycle 276. Unfortunately, Table HG 6 does not agree exactly for this cycle with the conversion of dates given by Mahler (1916, p. 574), but we have no other suggestion to offer.

HG 7. Tables for weekdays and movable feasts

There are two sub-tables, and they are not found in the *Almanach Perpetuum*. The first has the heading "Table for finding the weekday for [the first day of] January: epoch 1473"; MS Mu omits "epoch 1473."

According to Mahler's tables (Mahler 1916), December 31, 1472, was a Thursday; hence, January 1, 1473, was a Friday, as indicated in this table. Following a leap year of 366 days, 2 is added to the weekday of January 1; otherwise, 1 is added to the previous entry in the table.

The second sub-table has the heading "Table for finding the movable feasts of the Christians: the epoch of the 19-year cycle is 5226 [= 1465–66]" (MS L, f. 176r; MS Mu, f. 57b, with no heading; MS Ma, f. 5r).

This table has 35 rows and it is not arranged for a sequence of years; rather, it is ordered by the date on which the feasts occur in the Christian calendar. The epoch in the heading of MS Ma (f. 5r) is 1463, but MS Ma has another such table on f. 66v, giving 2 extra columns headed "rogationes april" and "corpus christi maius."

Table HG 7A: Weekday for the first day of January

year	weekday	[year AD]
1	6	1473
2	7	1474
3	1	1475
4	2	1476
5	4	1477
6	5	1478
.		
.		
26	2	1498
27	3	1499
28	4	1500

Table HG 7B: Movable Feasts

Year of the 19-year cycle	weekday for 1 Jan.	*cuaresma* Feb.	*pascua* Mar.	*Çinquesma*[a] May
13[b]	5	4	22	10
2[c]	4	5	23	11
0	3	6	24	12
10	2	7	25	13
.				
.				
Mar.			Apr.	June
0	6	10	25	13

a. (= Pentecost) This is probably a corruption of 'fiftieth' (perhaps from: *cincuagesima* or *quincuagesima*).
b. 16 in MS Ma
c. 5 in MS Ma

According to Mahler's tables (Mahler 1916), January 1, 1478 (= January 1 of year 13 of the 19-year cycle that began in September 1465) was a Thursday (= weekday 5), as in the first line of this table.

HG 8. "Table for finding the motion in corrected anomaly for each day from conjunction to opposition, or from opposition to conjunction"

Table HG 8A, columns I and II, is a transcription of Table 8 (MS L, f. 176r). Col. I represents the number of days after conjunction or opposition, and col. II the true daily motion in lunar anomaly, in degrees. Table HG 8 is not found in the *Almanach Perpetuum*. The entries in column II in this table are almost identical with those that can be derived from a table by

Judah ben Asher preserved in Vatican, MS Heb. 384, f. 362a-b, that gives the daily velocity in lunar anomaly as a function of double elongation at intervals of 1°. In order to compare the values in Zacut's table with those in Judah ben Asher's table, we first list in col. III the values for 2η taken from the Alfonsine Tables (Poulle 1984, p. 138) corresponding to the number of days in col. I. With the values in col. III as argument (rounded to degrees), we enter Judah ben Asher's table, and the results (rounded to minutes) are displayed in col. IV. The differences between cols. II and IV are displayed in col. V. We have recomputed Judah ben Asher's table directly from Ptolemy's complete lunar model, and the agreement is quite good. The purpose of Table HG 8, as well as that of the table of Judah ben Asher, is not clear, but it may have something to do with computing daily motions of the Moon. In any event, the two tables are closely related.

HG 9. Table for the daily lunar positions

This table is identical with Table AP 7 which displays the true longitudes of the Moon, given to minutes, and calculated for the meridian of Salamanca for noon of each day of a cycle of 31 years beginning on March 1, 1473. Note that MS Se has only 24 tables instead of 31.

HG 10. Table for the mean motion of lunar anomaly

This table is identical with Table AP 9 which displays the mean motion in lunar anomaly for each year in a period of 180 years. In MS Ma this table is found in a different set of tables, the *Tabule Verificate*.

HG 11. "Table for the correction of the anomaly"

This table is identical with Table AP 10 which gives the correction to be applied to the mean lunar anomaly to obtain the true lunar anomaly, i.e., the equation of center.

HG 12. Tables for the motion of the lunar nodes

The motion of the lunar nodes is presented in 2 tables: "Epoch 1473. Table for the motion of the true node for 93 years, and add to each of them [in later cycles ?] 1;15°" (MS L, f. 143v) and "Table for the motion of the mean node for 93 years, and subtract from each of them 1;15°" (MS L, f. 143r).

This table is identical with Table AP 11 (motion of the lunar nodes). The table in MS Ma appears in a different set of tables, the *Tabule Verificate*, and has March 1, 1461 as epoch.

HG 13. Table for lunar latitude

This table is identical with the first sub-table in Table AP 18 (for a maximum latitude of 5;0°).

Table HG 8A: Daily Motion in Lunar Anomaly

I	II	III	IV	V
	Zacut		Judah ben Asher	Diff. (II–IV)
(d)	(°)	(°)	(°)	(min)
1	16;38	24;23	16;38	0
2	16;35	48;46	16;35	0
3	16;13	73; 9	16;13	0
4	15;24	97;32	15;23	1
5	13;30	121;54	13;30	0
6	10;11	146;17	10;13	−2
7	6;14	170;40	6;12	2
8	4;37	195; 3	4;36	1
9	7; 7	219;26	7; 2	5
10	11; 2	243;49	11; 4	−2
11	14; 3	268;12	14; 1	2
12	15;40	292;35	15;40	0
13	16;21	316;58	16;18	3
14	16;35	341;20	16;35	0
15	16;39	5;43	16;39	0
16	16;37	30; 6	16;37	0
17	16;32	54;29	16;32	0
18	16; 6	78;52	16; 6	0
19	15; 2	103;15	15; 3	−1
20	12;56	127;37	12;53	3
21	9;14	152; 0	9;14	0
22	5;32	176;23	5;34	−2
23	4;51	200;46	4;51	0
24	8; 0	225;09	7;59	1
25	11;53	249;32	11;57	−4
26	14;31	273;55	14;31	0
27	15;53	298;18	15;52	1
28	16;26	322;41	16;26	0
29	16;36	347;03	16;35	1
30	16;40	11;26	16;40	0

HG 14. "Table for correcting the time of conjunction and opposition, and all aspects of the Moon with the planets"

The Lyon and Warsaw MSS have 3 additional columns for: 10;0°, 10;12°, and 10;24° (daily increment of elongation). It is the same as Table AP 17 (equation of syzygies).

HG 15. "Table for the motion of the true Sun from the true node"

In the sub-table for 56 years, MS L has "add for each revolution 3;32°." MS W only has the sub-table for 56 years. This table is the same as Table AP 12. In MS Ma the table has March 1, 1461 as epoch, and appears in a different set of tables, the *Tabule Verificate*.

HG 16. "Table for the parallax for the latitude of Salamanca which is 41;19°"

This table is identical with Table AP 14, except that MSS L, Ac, and Ma add a table for Capricorn, missing in the *Almanach Perpetuum*. In MS Ma this table is found in a different set of tables, the *Tabule Verificate*.

HG 17. "Table for correcting the eclipses and the parallax"

This table is identical with Table AP 13. The table in MS Ma has more entries, to two places, and appears in a different set of tables, the *Tabule Verificate*.

HG 18. "Tables for solar eclipses"

In spite of its title, this is a table for the lunar latitude, and it is identical with the second sub-table in Table AP 18 (for a maximum latitude of 4;29°). The table in MS Ma has twice as many entries, and appears in a different set of tables, the *Tabule Verificate*.

HG 19. "Table for solar eclipses at mean distance"

This table is identical with Table AP 15, but for a few entries; for example, the magnitude for an argument of 0;0° is 11;48d in MS L, but 11;47d in MSS Mu and Ac, and in the *Almanach Perpetuum*. The table in MS Ma has more than twice as many entries, and appears in a different set of tables, the *Tabule Verificate*.

HG 20. "Table for lunar eclipses at mean distance according to the opinion of R. Jacob Poel"

The words "according to the opinion of R. Jacob Poel" are missing in MS W. This table is the same as Table AP 16, except for the half-duration of the eclipse corresponding to an argument of 0;0°, where MSS L, Mu, W, and Ma have 1;49h instead of 1;45h as in Table AP 16. The table in MS Ma has more entries, and appears in a different set of tables, the *Tabule Verificate*.

HG 21. "Table for lunar eclipses at mean distance according to the opinion of R. Judah ben Asher"

This table is not found in the *Almanach Perpetuum*, but it is mentioned

Table HG 21A: Lunar Eclipses According to Judah ben Asher

lunar dist. from the node	**digits**	**half-dur. of eclipse**	**half-total. (min)**
11;30°	0	0; 0h	0
10;58	1	0;33	0
10;26	2	0;46	0
.			
.			
5; 6	12	1;37	0
4;34	13	1;44	22
.			
.			
0;50	20	1;49	48
0;18	21	1;49	49
0; 0	21;36	1;49	49

in chapter 7 of the *Ḥibbur* (Cantera 1931, pp. 180–183). A comparable table is found among the tables of Judah ben Verga (Paris, MS Heb. 1085, f. 97a; Oxford, MS Poc. 368 [Nb. 2044], f. 233a): in particular, the columns for the lunar distance from the node and the digits of eclipse are identical, but for copyist's errors; the entries in the other two columns are similar but systematically different. Judah ben Verga does not mention Judah ben Asher.

HG 22. "Table for finding the eclipsed area of the luminaries"

The 2 sub-tables displayed here are identical with the second sub-tables of Tables AP 15 and AP 16. In MS Ma they appear in a different set of tables, the *Tabule Verificate*.

HG 23. "Table for finding the ascendant and the 12 houses for latitude 41;19°"

This table is identical with Table AP 1. In MS Ma the 12 monthly sub-tables begin in December.

HG 24. "Table of arcs and sines"

The entries are given at 1°-intervals from 1° to 90°. This table is not found in the *Almanach Perpetuum*.

HG 25. "Table for finding the ascensions for all places"

The entries are given at 1°-intervals from 1° to 90°. MS Mu has another such table (f. 211a) with seconds in addition to degrees and minutes: see the selected entries in Table HG 25A, below. This table is not found in the *Almanach Perpetuum*.

The entries have been computed by taking the tangent of the argument, e.g., tan (30°) = 0;34,38, which is the entry in Table HG 25. The entry for 90° should not be 60, for tan (90°) is infinite. The table is to be used with the formula:

$$\sin \gamma = \tan \delta \cdot \tan \phi \qquad [1]$$

where γ is the difference between right and oblique ascension, δ is the declination, and ϕ is the geographical latitude (Goldstein 1967, p. 204).

HG 26. "Table of right ascensions for all places"

This table is identical with Table AP 19 which gives the normed right ascension, i.e., the right ascensions increased by 90°, as a function of ecliptic longitude.

HG 27. "Table for finding the oblique ascensions for all signs for the 7 climates and their midpoints"

The vertical argument is given at intervals of 10° of longitude, and the horizontal argument gives the longest daylight at intervals of 0;15h from 12;45h to 16;0h. This table is not found in the *Almanach Perpetuum*.

HG 28. Geographical coordinates

This is a list of cities with their geographical longitudes and latitudes; there are 3 columns with 32 entries each in MS L. This table corresponds to Table AP 49.

Table HG 25A: Ascensions

degree of lat. on Earth and the declin. of the star	diff. in horizon L and Mu, 164a	diff. in horizon Mu, 211a
1	0; 1	0; 1, 3
2	0; 2	0; 2, 6
.		
.		
30	0;34	0;34,38
60	1;44	1;43,55
70	2;45	2;44,53
80	5;40	5;40,10
90	60; 0	60; 0, 0

The first entries in MS L are:

1. Jerusalem	67;30[a]	33; 0
2. Egypt	64;51[b]	29;55
3. Babylon	80; 0[c]	33;30

a. Note in the margin of MS L: "in my opinion it is 66;30."
b. MSS Mu and Ac: 69;51.
c. Note in the margin of MS L: 78;0 according to R. Judah ben Asher [MSS Mu and Ac: 80;0].

At the bottom of the page in MS L there is a reference to Ptolemy's *Cosmographia* which is not found in MS Mu. Cohn 1918 (pp. 31–33) edited the table in MS Mu, and Cantera 1931 (pp. 363–370) edited the table as it appears in MSS L and Mu, comparing it with Table AP 49.

HG 29. "Table of the 120 conjunctions in the order given by R. Abraham Ibn ʿEzra in his *Book of the World* (*Sefer ha-ʿOlam*)"

This table lists the conjunctions of the planets taken 2, 3, 4, 5, 6, 7 at a time; it is just a list of the combinations. It is not found in the *Almanach Perpetuum*.

HG 30. "Table of the lunar eclipses for 50 years"
HG 31. "Table of the solar eclipses for 50 years"

These two tables correspond to Table AP 47, and are discussed in Chabás and Goldstein 1998. The Hebrew manuscripts have similar lists for eclipses, which differ slightly from the list in MS Se, ff. 6v-8r.

The title for HG 30 in the Hebrew version is "Table for lunar eclipses for 50 years, where years are counted from March" (MS L, f. 132r; MS Mu, f. 211a), and in it we find columns A, B (the weekday is given as a number such that Sunday = 1), C, D, G, and F, as listed below. Note that in the Hebrew version years are counted from March, whereas in MS Segovia they begin in January. Thus, the lunar eclipses occurring in Feb. 1487, Jan. 1497, and Feb. 1504 are listed there as Feb. 1486, Jan. 1496, and Feb. 1503, respectively. The Hebrew version agrees with MS Segovia, but for a few entries.

The title for HG 31 in the Hebrew version is "Table for solar eclipses for 50 years" (MS L, f. 132r; MS Mu, f. 183a), and in it we find columns A, B (the weekday is given as a number such that Sunday = 1), C (the heading is "true conjunction according to Poel"), C' (the heading is "apparent conjunction according to Poel"), G, and F. In the Hebrew version there is an extra column placed after (C) with the heading "true conjunction according to Alfonso," but it is restricted to a few values: 0;21h for 1478, 3;43h for 1481 (blank in MS Mu), 1;46h for 1485, 0;30h for 1487, 0;52h for 1491, and 1;13h for 1506. The Hebrew version has 1478 as the year of the first solar eclipse, and lists an additional eclipse for Jan. 23, 1524

Table HG 30A: Table of lunar eclipses in Segovia, MS 110

A		B		C	D	E	F	G	H	[N]
1475	Mar.[a]									
1475	Sep.	Thu.	14	18; 0	1;42	3;24	15;7[g]	0;36	1;12	4147
1476	Mar	Sun.	10	6;20	1;42	3;24	15	0;36	1;12	4148
1476	Sep.	Tue.	3	10;50	1;43	3;26	16	0;41	1;22	4149
1478	Jul.	Tue.	14	13;50	1; 7	2;14	5	0; 0	0; 0	4151
1479	Jul.	—[c]	3	15; 0	1;48	3;36	21	0;54	1;48	4153
1479	Jul.[b]	Tue.	28	11;40	1;42	3;24	15	0;36	1;12	4154
1482	Oct.	Sat.	26	4;28[f]	1;37	3;14	13	0;22	0;44	4157
1483	Apr.	Tue.	22	10; 0	1;44	3;28	17	0;45	1;30	4158
1483	Oct.	Wed.	15	12;17	1;47	3;34	19	0;50	1;40	4159
1487	Feb.	Wed.	7	14;40	1;43	3;26	16	0;41	1;22	4164
1489	Dec.	Mon.	7	16;22	1;43	3;26	16	0;41	1;22	4168
1490	Jun.	Wed.	2	9;40	1;49	3;38	23	0;55	1;50	4169
1490	Nov.	Fri.	26	16;45	1;46	3;32	18	0;48	1;36	4170
1493	Apr.	Mon.	1	13;11	1;43	3;26	16	0;41	1;22	4173
1494	Mar.	Fri.	21	13;40	1;43	3;26	16	0;41	1;22	4175
1494	Sep.	Mon.[d]	14	18;49	1;44	3;28	17	0;45	1;30	4176
1497	Jan.	Wed.	18	5;34	1;44	3;28	17	0;45	1;30	4178
1500	Nov.	Thu.	5	11;54	1;37	3;14	13	0;22	0;44	4182
1501	May.	Mon.	2	17;20	1;47	3;34	19	0;50	1;40	4184
1502	Oct.	Sat.	15	11;12	0;57	1;54	3[h]	0; 0	0; 0	4186
1504	Feb.	Thu.	29	12;30	1;43	3;26	16	0;41	1;22	4187
1505	Aug.	Thu.	14	7;24	1;42	3;24	15	0;38	1;16	4190
1508	Jun.	Tue.[e]	12	17;10	1;49	3;38	23	0;55	1;50	4194
1509	Jun.	Sat.	21	10;46	1;17	2;34	7	0; 0	0; 0	4196
1511	Oct.	Mon.	6	10;48	1;37	3;14	13	0;22	0;44	4199
1515	Jan.	Mon.	29	14;20	1;43	3;26	16	0;42[j]	1;24	4205
1516	Jan.	Sat.	19	5; 0	1;42	3;26	15;7[i]	0;38	1;16	4207
1516	Jul.	Sun.	13	10;50	1;40	3;20	14	0;30	1; 0	4208
1519	Nov.	Sun.	6	5; 0	1;48	3;36	20	0;52	1;44	4212
1522	Sep.	Fri.	5	11;22	1;42	3;24	15	0;36	1;12	4216
1523	Mar.	Sun.	1	7;30	1;44	3;28	17	0;45	1;30	4217
1523	Aug.	Tue.	25	14;17	1;44	3;28	17	0;45	1;30	4218

a. The rest of the line is blank. This line is missing in the Hebrew version.
b. Instead of Dec., as in the Hebrew version.
c. 7 (= Sat.) in the Hebrew version.
d. Instead of 1 (= Sun.), as in the Hebrew version.
e. Instead of 2 (= Mon.), as in the Hebrew version.
f. 4;30h in the Hebrew version.
g. 15 in the Hebrew version.
h. 4 in the Hebrew version.
i. 15 in the Hebrew version.
j. 0;43 in the Hebrew version.

Table HG 31A: Solar Eclipses in Segovia, MS 110

A		B		C		C'	F	G	H	[N]
1479[a]	Jul.	Wed.	29	0;51	29	0;55	4;30[k]	0;55	1;50	6383
1479	Dec.	Mon.[d]	13[g]	22;18	12	21;30	5; 0	0;47	1;34	6387
1481	May.	Mon.	28	4;28	28	5;53	4; 0	0;51[n]	1;42	6390
1482	May.	Fri.	17	5;30	17	7; 0	4; 0	0;41	1;22	6392
1485	Mar.	Wed.	16	2;20	16	3;46	10; 0	0;54	1;48	6398
1487	Jul.	Fri.	20	1;20	20	1;40	2; 0	0;30	1; 0	6403
1491	May.	Sun.	8	1;21	8	2;10	8; 0	0;52	1;44	6412
1492	Oct.	Sun.[e]	20	22;47	20	21;40	3; 0	0;37	1;14	6415
1493	Oct.	Thu.	10	1; 5	10	0;52	9; 0	0;54	1;48	6417
1502	Oct.	Sat.	0	19;53[h]	0	18;52[j]	7;30[l]	0;52	1;44	6438
1506	Jul.	Mon.	20	1;55	20	2;30	3; 0	0;37	1;14	6446
1512[b]	Mar.	Mon.	6	23;49[i]	7	0;25	4; 0	0;40	1;20	6462
1518	Jun.	Tue.[f]	7	18;22	7	17; 0	10;30[m]	0;55	1;50	6474
1525[c]										

a. Instead of 1478, as in the Hebrew version.
b. Instead of 1513, as in the Hebrew version.
c. The rest of the line is blank. In the Hebrew version there is a line here for the eclipse of Jan. 23, 1524 (where the year began in March).
d. Instead of 1 (= Sun.), as in the Hebrew version.
e. Instead of 7 (= Sat.), as in the Hebrew version.
f. Instead of 2 (= Mon.), as in the Hebrew version.
g. Instead of 12, as in the Hebrew version.
h. 1d 4;7h "before noon" in the Hebrew version.
i. 7d 0;11h "before noon" in the Hebrew version.
j. 5;36 "before noon" in the Hebrew version.
k. 12 in the Hebrew version.
l. 7 in the Hebrew version.
m. 10 in the Hebrew version.
n. 0;41 in the Hebrew version.

(Oppolzer No. 6489); this date corresponds to Jan. 23, 1525 in the year that begins in January.

There follows a transcription of the tables for lunar and solar eclipses as they appear in MS Se. For purposes of modern identification, in col. N we have added references to Oppolzer's canon of eclipses (1887). The headings in these tables are:

(A) Year-number
(B) Date (month, weekday, and day)
(C) Time of true syzygy (hours and minutes)
(C') Time of eclipse middle (day, hours and minutes)
(D) Half-duration (hours and minutes)
(E) Total duration (hours and minutes)
(F) Magnitude (digits and minutes of a digit)
(G) Half-duration of totality (hours and minutes)
(H) Duration of totality (hours and minutes)
(N) Oppolzer No.

HG 32. Table for declinations of the stars

There are 2 sub-tables which are not found in the *Almanach Perpetuum*, and we have not succeeded in explaining them. The heading of the first sub-table in MS L (f. 131v) is: "First table for finding the distance of each star from the equator, and also which degree is crossing the meridian."

Table HG 32A displays partially the first two columns (col. II has two subcolumns: IIa and IIb). The third column is headed: 0[s] 2;17//6[s] 2;17; the 4th is headed: 0[s] 4;35//6[s] 4;35, and so on. Note that the entry for 90 in col. IIa is 66;27°, i.e., 90° – 23;33°, indicating that this column presumably represents the complement of declination. But we do not understand the entries for 90° in the succeeding columns: (ivb): 67;29, declination 0;5, (vb): 68;32, declination 0;10, (vib): 69;38, declination 0;15, . . . The heading for col. I may refer to the argument of latitude counted from the northern limit (as Ptolemy has it).

The heading of the second sub-table in MS L (f. 130v) is: "Second Table for the declination from the equator and the degrees of mid-heaven."

Table HG 32B displays partially the first two columns (col. II has two subcolumns: IIa and IIb). The third column is headed: 0[s] 16;20//6[s] 16;20; the 4th is headed: 0[s] 18;48//6[s] 18;48; the 5th, 0[s] 21;18//6[s] 21;18; the 6th, 0[s] 23;52//6[s]23;52; and the 7th 0[s] 26;28//6[s] 26;28.

HG 33. Star list

This table is not found in MS Mu. The first 9 (of 61 entries) are displayed in Table HG 33A; we discuss this table together with Table AP 45, below.

HG 34. Longitude of Saturn

This table is the same as Table AP 20, given for 60 years. Note that MS L gives 62 years, and MSS Mu and Ac 61 years, whereas MS Se gives 30 years, and MS Ma 30 years and 3 months, beginning in January 1475.

HG 35. Center of Saturn

The 3 Hebrew MSS and MS Ac give the apogee of Saturn: 8s 12;54°. This table is the same as Table AP 21, given for 60 years. Note that MSS L and Mu add columns for years 61 and 62, and that MS Ma displays 59 years, beginning in January 1476. MS Ac also gives 62 years.

HG 36. Anomaly of Saturn

This table is the same as Table AP 22, given for 60 years. MSS L and Mu add columns for years 61 and 62, and MS Ma displays 59 years, beginning in January 1476.

Table HG 32A: Declination of the stars

[col. I] **degree of lat. of the star** **from the northern belt (?)**	**[col. II]** **0[s] 0; 0** **6[s] 0; 0**		. . .
	[col. IIa] **dist. from the** **equator: N**	**[col. IIb]** **mid-heaven** **Psc**	
0	0; 0	—	
2	1;50	29; 8	
4	3;40	28;15	
6	5;30	27;23	
.			
.			
		Aqu	
54	47;52	29; 0	
56	49;27	27; 6	
58	51; 0	25; 8	
60	52;33	22;58	
62	54; 3	20;40	
64	55;28	18;14	
66	56;51	15;40	
.			
.			
76		Cap	
.			
.			
87	66; 9	7; 0	
88	66;23	4;25	
89	66;25	2;20	
90	66;27	0; 0	
	declin.* 0; 0		

* On fol. 130r, at the bottom there is a heading for this row (below the 90) that seems to mean parallax (*ḥilluf*: *lit.* 'difference') in declination on account of the ascensions(?).

HG 37. Latitude of Saturn

This table is the same as Table AP 23. MSS L and Mu read "Its ascending node is 140° before its apogee," and MS Ac adds "and 40° after its apogee."

HG 38. Longitude of Jupiter

This table is the same as Table AP 24, given for 85 years. Note that MSS L and Ac also give 85 years, but MSS Ma and Se display 83 years, beginning in March 1475.

Table HG 32B: Declination of the stars

[col. I] Lat. from the belt to the North	[col. II] 0[s] 13;55 6[s] 13;55		. . .
	[col. IIa] Decl. from the equator to the North	[col. IIb] mid-heaven Aries	
0	6; 0	—	
2	7;51	13; 5	
4	9;42	12;14	
6	11;32	11;24	
8	13;23	10;33	
10	15;13	9;41	
.			
.			
32		Psc	
.			
.			
56	55;51	11;26	
58	57;25	9;25	
60	58;58	7;10	
62	60;28	4;47	
64	61;55	2;15	
66		Aqr	
.			
.			
83		Cap	
.			
.			
88	73; 5	17; 0	
89	73; 7	14;23	
90	73;11	11;24	

HG 39. "Table for finding the rulerships of the 7 planets in the 12 signs" (MS L, f. 94r—written in a different hand; MS Se, f. 84v)

This table corresponds to the tables in MS Mu, f. 218a-b. It is not found in the *Almanach Perpetuum*. A very similar table is found in London, MS Sassoon 823, pp. 215–216.

There follows a set of astrological indications arranged by astrological house, and then in a separate paragraph:

Table HG 33A: Star list (L, f. 114v-r; W, f. 61a-b) All entries for magnitude longitude, latitude, and associated planets agree with Table AP 45, unless otherwise noted.

[No.]	Mag.	Name	Longitude		Lat.		Assoc. Plan.	[Mod.]	
1.	1	sof ha-nahar	Ari	6;48	13;30	S[a]	Jup	θ	Eri
2.	1[b]	rosh ha-gol	Tau	6;18	23; 0	N	Mars/Merc	β	Per
3.	1	ʿeyn ha-shor ha-yemini[c]	Tau	19;18	5;10	S	Mars	α	Tau
4.	1	ketef nose’ ha-resen[d]	Gem	1;38	22;30	N	Jup/Merc	α	Aur
5.	1	ketef ha-gibbor	Gem	8;38	17;30	S	Mars/Merc	α	Ori
6.	1	regel ha-gibbor	Tau	26;28	31;30	S	Sat/Jup	β	Ori
7.	1	qeṣeh ha-sefina[e]	Gem	23;48	75;40	S	Sat	α	Car
8.	1	ha-kelev ha-gadol al-ʿabur[f]	Gem	24;18	39;10	S	Jup/Mars	α	CMa
9.	1	ha-kelev ha-qaṭan al-gumayṣa[g]	Cnc	5;48	16;10	S	Merc/ a little Mars	α	CMi
.									
.									

a. W: 53;30 S; L mg.: “another copy: 53 [instead of 13], and it is more correct.” Table AP 45 has 61;30 S; *Almagest* has 53;30 S, often misread in Arabic, and derivative texts in Latin and Hebrew, as 13;30 S (Goldstein and Chabás 1996, p. 323).
b. L mg. and W mg.: “another copy: 2”. Table AP 45 has 1; Ibn al-Kammād’s list has 2 (Goldstein and Chabás 1996, p. 321).
c. L mg. and W mg.: “aldabaran”. In Ptolemy’s star catalogue this is the star on the southern eye of the bull; hence, “yemini” here means “southern”, rather than “right-[side]”; cf. Arabic “yamanī” that can mean “southern” or “right-[side]”.
d. W: “ketef nose’ sarṭan”; L mg. and W mg.: “al-ʿayyuq”.
e. L mg. and W mg.: “suhayl”.
f. L mg.: “shiʿrat al-yamaniya”: Nallino 1899–1907, 2:179.
g. L mg.: “shiʿrat al-sha’amiya”: Nallino 1899–1907, 2:179.

> The houses of joy: Mercury 1st; Moon 3rd; Venus 5th; Mars 6th; Sun 9th; Jupiter 11th; Saturn 12th. The Sun and the Moon do not have terms, but the Sun rules from the beginning of Leo to the end of Capricorn, and the Moon rules from the beginning of Aquarius to the end of Cancer. . . .

Note that the houses of joy here agree with those listed below the table in MS Mu, f. 218a, except for Mars (8th house instead of 6th house). For the rulerships of the Sun and the Moon, see Bīrūnī, *Tafhīm*, (Wright 1934, p. 256).

The entries in all the columns are represented in MS Mu, f. 218a–b, but there are additional columns for 'joys' (here associating planets with zodiacal signs) and for 'nocturnal triplicities' in MS Mu, f. 218a, that have been omitted in Table HG 39A: see Tables HG 70A and 71A, and their notes, for these data. Moreover, MS Mu, f. 218a, has no word in the heading corresponding to 'dignity' ('Dign.', in Table HG 39A). The dignities given here agree with Bīrūnī, *Tafhīm*, trans. Wright 1934, p. 306, except that for al-Bīrūnī 'terms' have dignity 3 and triplicities have dignity 2. In col. III the planetary exaltations are given with the degree within the zodiacal sign, whereas in MS Mu, f. 218a, the degrees are omitted.

In Ptolemy's *Tetrabiblos* there are two systems of 'terms,' an unequal fivefold division of each zodiacal sign, where each of its five terms is ruled by a different planet (not including the Sun and the Moon). MS L, f. 94r, has the list of terms according to the Egyptians (*Tetrabiblos*, I, 20; ed. Robbins 1940, p. 97), although the heading simply has 'terms'. MS Mu, f. 218b, has the list according to the Egyptians as well as the list according to Ptolemy (*Tetrabiblos*, I, 21; ed. Robbins 1940, p. 107), and they are so labeled in the heading. These lists have been omitted in Table HG 39A.

HG 40. "Table of visibility for the 5 planets in the 5th [*sic*] climate"

This table appears in MS L (f. 93v), and in MS Mu (f. 215a: superior planets; f. 159a: inferior planets). See also Table HG 60. Despite the heading, the

Table HG 39A: The rulerships of the 7 planets

I	II	III	IV	V	VI
[Dign.]	5	4	3	1	2
Zod. sign	House	Exalt.	Triplic.	Faces	Terms
Ari	Mars	Sun 19	Sun/Jup/Sat	Mars/Sun/Ven	
Tau	Ven	Moon 3	Ven/Moon/Mars	Merc/Moon/Sat	
.					
.					
Psc	Jup	Ven 27	Ven/Mars/Moon	Sat/Jup/Mars	

entries are for the 4th climate, as found in Ptolemy's *Handy Tables,* al-Battānī's zij, and the Toledan Tables (Toomer 1968, pp. 73 ff). The table in MS Ma, f. 66r, is very much the same, but note that the headings of the columns for Jupiter and Mars have been interchanged. There is no table in the *Almanach Perpetuum* that corresponds to this one.

HG 41. Center of Jupiter

The 3 Hebrew MSS give the apogee of Jupiter: 5s 23;7°. This table is the same as Table AP 25, given for 84 years. Note that MS L adds columns for years 85 and 86, and that MS Ma displays 83 years, beginning in January 1476. MS Ac also gives 86 years.

HG 42. Anomaly of Jupiter

This table begins in January, and it is the same as Table AP 26, given for 86 years. Note that MS Ma displays 83 years, beginning in January 1476.

HG 43. Latitude of Jupiter

MSS L, Mu, and Ac read "Its ascending node is 70° before its apogee." This table is the same as Table AP 27.

HG 44. Longitude of Mars

This table begins in January, and it is the same as Table AP 28, given for 80 years. Note that MSS L and Mu add a column for year 81. MS Ac also gives 81 years, beginning in March 1473. MS Ma displays 79 years, beginning in March 1475, and MS Se just 32 years, also beginning in March 1475.

HG 45. Center of Mars

The 3 Hebrew MSS give the apogee of Mars: 4s 14;42°. This table begins in January, and it is the same as Table AP 30, given for 34 years. Note that MSS L, Mu, and Ac add a column for year 35, and that MS Ma displays 32 years, beginning in January 1476.

HG 46. Anomaly of Mars

This table begins in January, and it is the same as Table AP 31, given for 35 years. Note that the last column in MS W is for year 32, and that MS Ma displays 32 years, beginning in January 1476.

HG 47. Latitude of Mars

MSS L, Mu, and Ac read "Its ascending node is 90° before its apogee." This table is the same as Table AP 32.

HG 48. Correction of Mars

This table is the same as Table AP 29.

HG 49. Longitude of Venus

This table begins in March, and it is the same as Table AP 33, given for 8 years. MSS Ac and Ma display 8 years, beginning in March 1473. Surprisingly, MS Se does not display the daily positions of Venus, but its positions at intervals of 5 days for a period of 32 years, also beginning in March 1473.

HG 50. "Table for correcting Venus in other revolutions" and "Table for finding the anomaly in all revolutions"

Note that the word translated as 'revolutions' (*sibbuvim*) refers to cycles of 8 years in this case, whereas the term 'cycle' (*maḥzor*) is restricted to calendrical cycles. These two sub-tables are identical with Table AP 34 (correction of Venus). MS Se only gives the first sub-table, where the entries are given in days, minutes, and hours.

HG 51. Center of the Sun and Venus

The 3 Hebrew MSS and MS Ac give the apogee: 3s 0;56°. This table begins in January, and it is the same as Table AP 35, given for 4 years. Note that MS Ma displays 4 years, beginning in January 1476.

HG 52. Anomaly of Venus

A note above the table in MSS L and Mu (f. 140a) reads "for each revolution add 1;27 1/2°," whereas MS Ac reads "add 27 min 30 s."

This table begins in January, and it is the same as Table AP 36, given for 9 years. Note that in MS L there are columns for only 8 years, whereas MSS Mu and Ac display 11 years, and that MS Ma displays 8 years, beginning in January 1476.

HG 53. Latitude of Venus

In MS L there is a note above the first table: "[Belonging] to me, Abraham Zacut. First table for finding the latitude of Venus. . . ." This table is identical with Table AP 37.

HG 54. Longitude of Mercury

This table begins in March, and it is the same as Table AP 38, given for 125 years. Note that MS Ma displays 20 years, and MS Se 31 years, beginning in March 1475 in both cases.

HG 55. Correction of Mercury

This table is the same as Table AP 39.

HG 56. Center of Mercury

The two Hebrew MSS and MS Ac give the apogee: 7s 0;10°. This table begins in January, and it is the same as Table AP 40, given for 4 years. Note that MS Ma displays 4 years, beginning in January 1476.

MS Ma has yet another table for the same purpose on f. 66r, the heading of which is "Tabula cursus centri mercurii in quodlibet die," with entries at intervals of 4 days (the entries in Table AP 40 are given at 10-day intervals) from March to December, and the part of the table intended for January and February is blank. This is not the case in MS Se, where we find exactly the same table, including the entries for January and February, under the title "Tabula motus centri mercurii" (f. 82r). The first entry in both tables is 11s 18;49° and corresponds to March 4.

HG 57. Anomaly of Mercury

This table begins in January, and it is the same as Table AP 41, given for 40 years. Note that MSS L, Mu, and Ac have columns up to year 43, and that MS Ma displays 40 years, beginning in January 1476.

HG 58. Unequal motion of Mercury

This table is the same as Table AP 42, except for slight differences.

HG 59. Latitude of Mercury

This table is the same as Table AP 43. MS L adds a column for 12s 0°.

HG 60. Visibility table for Venus and Mercury

This table is found in MS Mu in a hand that is more cursive than that of the rest of the manuscript, and corresponds to part of HG 40 in MS L; it is not found in the *Almanach Perpetuum*.

HG 61. First and second stations

The headings for the columns in MS L read "corrected anomaly," and those for the rows read "corrected center," and are given at 6° intervals. This table is not found in the *Almanach Perpetuum*.

HG 62. Sexagesimal multiplication table

This table is the same as Table AP 44.

Table HG 63A: Number of days in a cycle

cycle	No. of days	
1	0,7,21,10 :	19 years
.		
.		
320	:	6080 years

Table HG 63B: Number of days in a month

simple months	ordinary	[years] defective	full
1	1, 0	1, 0	1, 0
2	1,29	1,29	2, 0
3	2,29	2,28	3, 0
4	3,28	3,27	3,29
5	4,28	4,27	4,29
.			
.			
12	11,24	11,23	11,25
[year with a] leap month			
1	1, 0	1, 0	1, 0
2	1,29	1,29	2, 0
3	2,29	2,28	3, 0
4	3,28	3,27	3,29
5	4,28	4,27	4,29
.			
.			
13	12,24	12,23	12,25

HG 63. "Table for finding the number of [days corresponding to] cycles, years, and days [read: months] of the [Era of] Creation"

This table (together with its sub-tables) gives the number of days (in base 30), in 19-year cycles from 1 to 320, in years from 1 to 19, and in months for the various year lengths in the Jewish calendar. According to the sub-tables for the various year lengths, the months 2 and 3 (Marḥeshvan and Kislev) are sometimes 29 days and sometimes 30 days, in agreement with other Hebrew sources. However, this does not conform to what is said in chapter 18 of the *Ḥibbur* (Cantera 1931, p. 342), where we are told that months 3 and 4 (Kislev and Ṭevet) vary in length. Note that number '19', and subsequent multiples of it, are in decimal place-value notation using the first 9 letters of the Hebrew alphabet plus a symbol for zero, as described in chapter 18 of the *Ḥibbur*. This table is not found in the *Almanach Perpetuum*.

Note that there is no difference between the two sub-tables in Table HG 63B for months 1 to 5, and that months 2 and 3 are sometimes 29 days and sometimes 30 days; all other months have a fixed duration.

HG 64. "Table for finding the conjunctions of the Moon" (L adds: "according to our Sages")

There are 6 sub-tables for cycles 1 to 320, years 1 to 19, months 1 to 12, days 1 to 7, hours 1 to 24, and parts 1 to 1080 (1 hour = 1080 parts). They are mentioned in chapter 18 of the *Ḥibbur*. This table is not found in the *Almanach Perpetuum*.

The radix for cycle 1 of the Jewish calendar (Era of Creation) is (2) 5h 204p, as in Gandz, Obermann, and Neugebauer 1956, p. 116, and the length of the 19 year cycle is (2) 16h 595p, as in Gandz, Obermann, and Neugebauer, 1956, p. 115. Note that, following Neugebauer's notation, (2) means weekday 2, i.e., Monday.

Tables HG 64A: Conjunctions of the Moon

cycle	day	hour	part
1	(2)	5	204[a]
2	(4)	21	799
3	(7)	14[b]	314
4	(3)	6	909
5	(5)	23	424
.			
.			
320	(6)	4	1009

a. with L; Mu: 24
b. with L; Mu: 4

Table HG 64B: Conjunctions of the Moon (years)

year	day	hour	part
1	(4)	8	976
2	(1)	17	672
.			
.			
19	(2)	16	595

Table HG 64C: Conjunctions of the Moon (months)

month	day	hour	part
1	(1)	12	793
2	(3)	1	506
.			
.			
12	(4)	8	876
half-month	(0)[a]	18	396½

a. With Mu; L: (2)

Table HG 64D: Conjunctions of the Moon (days)

day	cycle	month
1	220	140
2	441	45
3	661	185
4	110	70[a]
5	330	210
6	551	115
7	772	20

a. read 90?

Table HG 64E: Conjunctions of the Moon (hours)

hour	cycle	month
1	556	20
2	340	20
3	124	20
4	680	40
5	464	40
6	248	40
.		
.		
24	220	140

Table HG 64F: Conjunctions of the Moon (parts)

part	cycle	month
1	316	117
2	632	234
3	177	96
4	493	213
5	38	75
6	354	192
7	671	74[a]
8	215	171
9	532	53[b]
10	76	150
20	153	65[c]
30	229	215
40	306	130
50	383	45
60	459	195
70	536	110
80	613	25
90	689	175
100	766	90
200	760	160 (diff = 70)
300	756	230
.		
.		
1000	715	15
1080	556	20

a. read 54?
b. read 33?
c. read 45?

Note that 319 · [2d 16h 595p] + (2) 5h 204p = (3) 23h 805p + (2) 5h 595p = (6) 4h 1009p, the entry for 320 cycles. The sub-table for the years reflect the fact that years are not all of the same length in the Jewish calendar. Note also that 235 · [(1) 12h 793p] = (2) 16h 595p, the entry for 19 years [= 235 months]. Thus, the tables for cycles, years, and months are mutually consistent.

In the sub-table for the days, there seems to be an error in the months for day 4, but the subsequent entries agree with the 'wrong' value. If 7d corresponded to 2 · 772 + 40/235, then 1d = 220 + 140/235. We do not know the significance of this sub-table or of the following two sub-tables.

In the sub-table for the hours, the line-by-line difference in the cycles is 556 mod 772. Now, 556 · 24 = 220 (mod 772), and the entry for 24h equals the entry for 1d. The line-by-line difference in months is always either 0 or 20, and the entries increase from 20 to 140.

In the last sub-table, the differences beginning with argument 10 are 150 (mod. 235), rather than mod. 255 which predominated earlier (with some differences mod. 235).

HG 65. "Table for finding the *sigla* for each year of the cycle from the date of the first conjunction (*molad*) of the cycle"

This system for the Jewish calendar is described in the canons to the *Ḥibbur*, chap. 18. A discussion of the contents of this table is found in Mahler (1916, pp. 499 ff). Note that the term *molad* does not refer either to the true or mean conjunction of the Sun and the Moon as calculated from astronomical tables. Rather it is a calendrical convention according to which the time difference between one *molad* and the next is constant, 29d 12h 793p (= 29;31,50,8,20d, the standard value for the length of the synodic month) and, in general, the first day of the calendrical month is the day of the *molad*, unless a rule for postponing the first day of the month applies: see Maimonides, *Sanctification of the New Moon*, trans. Gandz 1956, p. 32. So the interval from one *molad* to the next is equal to the mean synodic month, but the *molad* is closer to the time of the appearance of the new crescent than is mean conjunction (as usually defined astronomically).

Note that the entries in the unnumbered line after line 61 repeat the entries in line 1. Note also that *s* refers to a simple year of 12 months,

Table HG 65A: *Sigla* for each year of the cycle

	limits of the *molad*				years				
mg*	No.	week-day	h	parts	1(s)	2(s)	3(b)	. . .	19(b)
	1	1	0	408	BHG	EXA	GKZ	. . .	EXG
1	2	1	5	333	BHG	EXA	GKZ	. . .	EXG
	3	1	7	858	BHG	EXA	GKZ	. . .	EXG
.									
.									
2	9	2	0	408	BXE	ZXG	EHA	. . .	ZHG
.									
.									
	31	4	14	175	EKZ	BXE	ZHG	. . .	BHE
.									
.									
	61	7	20	560	BHG	EKZ	BXZ	. . .	EXG
	—	1	0	408	BHG	EXA	GKZ	. . .	EXG

* The numbers in the margin only appear in L on f. 27r where they emphasize the weekday in the column under "limits of the *molad*."

and *b* to a leap year of 13 months. We have used E to transcribe the letter *he*, H for the letter *ḥet*, and X for the letter *shin*. According to chapter 18 of the *Ḥibbur* (MS B, f. 50b; Cantera 1931, p. 355), the middle letter is always H (for 'defective'), K (for 'ordinary'), or X (for 'full'). A fuller explanation is given in the canons earlier in the same chapter (MS B, f. 46a; Cantera 1931, p. 342). The first letter refers to the first day of the month, Tishri. The second letter refers to the months, Kislev and Ṭevet: if they are both 'defective', the year is called 'defective', if they are both 'full', the year is called 'full', and if they are 'ordinary' (i.e., one 'full' and one 'defective'), the year is called 'ordinary' (see Maimonides, *Sanctification of the New Moon*, trans. Gandz 1956, pp. 34 ff, where the rule is stated for the months, Marḥeshvan and Kislev, rather than for the months, Kislev and Ṭevet. Maimonides's rule is also found in Mahler (1916, p. 483). Note that a 'defective' month is 29 days and a 'full' month is 30 days. The third letter refers to the first day of the month, Nisan.

The first day of Tishri cannot fall on a Sunday, Wednesday, or Friday (i.e., weekdays 1, 4, or 6: in alphabetical notation: A, D, W). If the *molad* falls on one of these days, the first day of the month is postponed to the next day: Maimonides, *Sanctification of the New Moon*, trans. Gandz 1956, pp. 30 ff. Hence, the only letters that can appear in the first place of the three letter *sigla* are: B (2), G (3), E (5), Z (7).

3.2 Other tables related to the *Ḥibbur*

3.2.1 MS L

f. 25r. "Table for finding the number of years and days of the Christians"

This table is for converting a date in the Christian calendar to a number of days, base 30, e.g. 1000 years = 13,15,25,0d (up to 8000 years).

f. 24v. A similar table for converting a date in the Hijra calendar to a number of days, base 30 (up to 9000 years).

f. 24r. A similar table for converting a date in the Persian calendar to a number of days, base 30.

f. 23v. (1) Difference in days between eras (base 30).

f. 23v. (2) [Planetary] radices for the beginning of the Incarnation, the last day of December, according to Alfonso for the longitude of Toledo.

f. 23v. (3) [Planetary] radices for the last day of May of the year 1252 for the longitude of Toledo, the day when Don Alfonso became king.

Table HG 66A: Comparison between the radices for 1252 in MS L and the Alfonsine Tables (1483)

	MS L (f. 23v)		Alfonsine Tables	
Mean Saturn	8s	24;22, 8, 9	8s	24;22, 8,11
Apog. Saturn	8	2;17,20, 0	8	2;17,20, 5
Mean Jupiter	0	16; 8, 3,22	0	16; 8, 3,24
Apog. Jupiter	5	12;24, 8,15	5	12;24, 9,20
Mean Mars	6	1;25,24, 0	6	1;25,24, 2
Apog. Mars	4	4;11,20, 7	4	4;11,27,28
Mean Sun	2	16;18,18, 7	2	16;18,18, 5
Apog. Sun	2	20;18, 0, 0	2	20;18,15, 0
Anom. Venus	1	15;22,28,22	1	15;22,28,25
Anom. Mercury	7	3;24, 9,22	7	3;24, 9,22
Apog. Mercury	6	19;25,18, 0	6	19;25,17,28
Mean Moon	11	6; 2,20, 7	11	6; 2,20, 9
Anom. Moon	8	10;25,25, 0	8	10;25,25, 1
Asc. Node	5	26; 6,11,15	5	26; 6,11,16

All numbers are given in base 30. They are listed in Table HG 66A, together with the radices appearing in the Alfonsine Tables (ed. 1483) translated from base 60 to base 30.

Except for the apogee of Mars, where the discrepancy might be due to a copyist's error in MS L, there is very good agreement.

f. 23r. Epoch 1401 of the Incarnation, noon, before Jan. Above the table: radices of the planets for Salamanca whose longitude from the west is 25;46° — this heading applies to the entire set that follows; all entries are displayed in base 30 notation. (MS Ac, f. 118r, gives 21;46° for the longitude of Salamanca.)

Note that the epoch 1401 is mentioned in chapter 19 of the *Ḥibbur* (Cantera 1931, pp. 234, 358): *comienço del año de .1401. de los naçarenos . . . e uan fundadas sobre longura de .25. grados e 46. mjnutos de occidente*, where 25;46° is the longitude assigned to Salamanca.

Comparing the radices for 1401 in MS L with those computed with the Alfonsine Tables for noon of December 31, 1400 (converted to base 30) we find good agreement for the mean motions of the planets, but a systematic difference of about 0;5° (base 30) between the entries in MS L and the computed values for the apogees (see Table HG 67A).

(1) Table for the mean motion of the Sun: radix 9s 18;19,1°; apogee 3s 0;11,26,17°; entries to 9 places, base 30: days 1 to 30.

(2) Table for the mean Moon: radix 3s 22;9,4,28°; entries base 30.

Table HG 67A: The radices for 1401 in MS L

	MS L (ff. 23r-21r) (base 30)	Computed (base 60)	Computed (base 30)
Mean Sun	9s 18;19, 9, 1	4s 48;38, 9	9s 18;19, 2
Apog. Sun	3s 0;11,26,17	1s 30;13,30	3s 0; 6,22
Mean Moon	3s 22; 9, 4,28	1s 52;12,20	3s 22; 6, 5
Mean lunar anom.	3s 10;28,10,14,0[a]		
Asc. node	5s 20; 2,24,20		
Mean Saturn	9s 12;19, 9,21	4s 42;38,38	9s 12;19, 9
Apog. Saturn	8s 12;11, 1,10	4s 12;11,51	8s 12; 2,28
Mean Jupiter	6s 27;19,25,29	3s 27;39,41	6s 27;19,25
Apog. Jupiter	5s 22;17, 2,25[b]	2s 52;25, 9	5s 22;12,17
Mean Mars	6s 3;16,18, 4	3s 3;32,58	6s 3;16,14
Apog. Mars	4s 14; 5, 9, 2	2s 14; 0,22	4s 14; 0, 5
Anom. Venus	0s 24;20,26, 6		
Anom. Mercury	11s 7;20,10,22		
Apog. Mercury	6s 29;18,29, 2[c]	3s 29;27,42	6s 29;13,25

a. MS Ac has 3s 10;28,14, 0
b. MS Ac has 5s 22;17,20,25
c. MS Ac has 6s 19;18,29, 2

f. 22v.
(1) Table for the lunar anomaly: radix 3s 10;28,10(?),14,0°.
(2) Table for the lunar node: radix 5s 20;2,24,2°.

f. 22r.
(1) Table for the mean motion of Saturn: radix 9s 12;19,9,21°; apogee 8s 12;11,1,10°.
(2) Table for the mean motion of Jupiter: radix 6s 27;19,25,29°; apogee 5s 22;17,2,25°.

f. 21v.
(1) Table for the mean motion of Mars: radix 6s 3;16,18,4°; apogee 4s 14;5,9,2°.
(2) The mean motion of Venus and its apogee are the same as those for the Sun; anomaly of Venus: radix 0s 24;20,26,6°.

f. 21r.
(1) Table for the anomaly of Mercury: radix 11s 7;20,10,22°; apogee 6s 29;18,29,2°.
(2) Table for the motion of the apogees for all planets except the Moon: days from 1 to 30. The entry for 30d is 0;0,16,8,23,21,6° (base 30) which corresponds to 0;0,1,1,5,55,36° (base 60) or 1° in about 290 Julian years.

A value very close to this appears in the zij of Ibn al-Kammād for the motion of the solar apogee (Chabás and Goldstein 1994, p. 28). The idea that the motion of the solar apogee is to be applied as well to the planetary apogees is reported by Ibn al-Kammād, Ibn Isḥāq (fl. ca. 1193–1222), and Ibn al-Hā'im (fl. 1205), all three probably reflecting the view of Azarquiel (Mestres 1996, p. 394; Samsó and Millás 1998, pp. 268–270). In Vatican, MS Heb. 384, there are values for the planetary apogees that can be derived from those in Ibn al-Kammād's zij by applying the motion of the solar apogee to all of them (Goldstein 1998, p. 184 n. 22). However, in the Alfonsine tradition the planetary apogees do not move with the solar apogee. Hence, in this case, Zacut has departed from the Alfonsine tradition to which he generally adheres.

ff. 20v-19v. Tables related to the Jewish calendar.

On ff. 20r-19v there are two tables for the *tequfot* (i.e., solstices and equinoxes), one according to R. Adda (top of both pages: see also MS Mu, 193a–193b), and another according to Samuel (bottom of both pages). This topic is discussed by Maimonides (see Gandz et al. 1956, pp. 36–42, 94–98; commentary by Neugebauer, pp. 113–123); and in Mahler 1916, pp. 159 ff, 506 ff. The canons of the *Ḥibbur* (MS S, f. 56r; Cantera 1931, p. 227) mention two Talmudic rabbis, Rabada [= R. Adda bar Ahaba] and Samuel, who were associated in the Jewish tradition with calendrical matters, particularly with the *tequfot.* The table ascribed to Samuel is based on a solar year of 365 1/4 days and all four seasons have the same length, 91d 7;30h; whereas the table ascribed to R. Adda is based on a somewhat smaller solar year to conform with the 19-year cycle (where 19y = 235 · 29;31,50,8,20d), but the seasons are still assumed to be of equal length. Thus, the level of astronomical knowledge reflected in these tables is well below that of Zacut's Table HG 5 (= Table AP 6) which yields better values for the length of the seasons.

The corresponding tables in MS Ac begin on f. 115r. It is unusual to find these tables in a Latin manuscript, whereas they frequently appear in Hebrew texts. In fact, MS Ac is the only Latin copy of the *Ḥibbur* with these tables, indicating that the translator respected the Jewish context into which Zacut inserted his astronomy.

f. 19r. A brief text containing instructions for the use of tables.

f. 18v. Table for correcting the centers of the planets, and minutes of proportion. (MS Mu, f. 194a)

The entries are displayed at 6° intervals of the argument of center, and they agree with those in the Alfonsine Tables (ed. 1483).

f.18r.
(1) Table for the correction due to the corrected anomaly for the 3 superior planets.
(2) Table for the correction due to the anomaly of the 3 planets below the Sun.

ff. 17v-2v. Star catalogue. (MS Mu, ff. 195a-210a, epoch 1478; f. 210b blank)

The heading is: "Table for finding the positions of the fixed stars in the 9th orb, and their latitudes from the belt [of the ecliptic] for this year, 1478." The entries are arranged by constellation, with numbers and names for each star within the constellation, longitude, latitude, magnitude, and direction of latitude (N or S). The notation for longitude is given in 'natural' signs, each of which is divided into 30 degrees, and each degree into 60 minutes; the latitudes are the same as in the Alfonsine Tables (ed. 1483).

The terminology is not the same as in the star catalogue in the Hebrew translation of the Alfonsine Tables (Munich, MS Heb. 126, where this list has the heading: "Table for the positions of the bright fixed stars of first and second magnitude for the beginning of year [1]496"), and in that translation the coordinates are the same as in the *editio princeps* (with 'physical' signs of 60°); the longitudes that we checked in Munich, MS Heb. 126 were identical with those in the Alfonsine Tables (ed. 1483), i.e., precession was not applied to these longitudes.

The first entry for UMi has coordinates 2s 20;40°, 66;0° N; the second entry has coordinates 2s 23;0°, 70° N. The first entry for UMa has coordinates 3s 15;50°, 39;50° N. In all cases the latitudes agree with the Alfonsine Tables (ed. 1483). The precession in all three cases is 3;12°. In 226 years (= 1478 – 1252), a precession of 3;12° corresponds to 1° in 70;37,30 years.

f. 2r. "Table for correcting Mars for other revolutions, and it belongs earlier" (MS W, f. 104a)

This table gives the correction to be added to the longitude of Mars after one cycle of 79 years, and it is the same as Table AP 29.

f. 1v-r. Corresponding dates in the Jewish calendar with those in the Christian calendar, beginning with year [5]237 and ending with [5]260 A.M. (= 1476/77 to 1499/1500).

3.2.2 MS Mu

On f. 193a-b there is a table with the heading, "Table of *Tequfot* according to R. Adda." See the description of MS L, ff. 20r-19v, above.

Table HG 68A: Animodar

degree of the Moon from the asc.	eastern duration (ʿ*ikkuv*)
61	277d 20;24h
62	277d 22;13
63	278d 0; 2
.	
.	
179	286d 19;25
180	286d 21;15

Folios 215b-216a, following the table for the visibility of the planets (Table HG 40), are blank. The next table is on f. 217a, and has no heading. It is an animodar table similar to Table AP 46.

The argument in this table (see Table HG 68A) is the argument of lunar latitude, given for all integer degrees from 61° to 180°. The entries are given in days, hours, and minutes, and correspond to the entries in the column headed *Tempus more orientalis* of Table AP 46 (here given in weeks, days, hours, and minutes). Note that consecutive entries differ by 1;49h or 1;50h in a sequence such that, usually, entries corresponding to arguments 3° apart differ by 5;28°. For the origin of the term "animodar" and other such tables, see comments to Table AP 46.

The heading for the next table (f. 217b) is: "Table in signs, degrees, and minutes in which the Moon moves according to its mean motion during the time the child is in his mother's womb from its [the Moon's] place at the time of conception to its place at the time of birth"; this table gives the mean motions of the Moon in longitude and anomaly according to the duration of pregnancy from 259d to 287d (see Table HG 69A).

Note that for the motion of the Moon 259d · 13;10°/d is about 3410°. Hence the entry for 259d corresponds to 3412;40° (= 9 · 360° + 172;40°), and 3412;40°/259d = 13;10,34,45°/d. Similarly, 287d · 13;10°/d = 3778;50°. Hence the entry for 287d is to be understood as 3780° (=

Table HG 69A: Motions of the Moon during pregnancy

days that the child is in the womb	mean motion of the Moon	days that the child is in the womb	mean motion of the Moon in anomaly
259d	5s 22;40°	259d	4s 23;50°
.			
.			
287d	6s 0; 0°	287d	4s 28;56°

360° · 10 + 180°), and 3780°/287d = 13;10,14,38°/d. For the motion in anomaly 259d · 13;3°/d is about 3380°. Hence the entry for 259d is to be understood as 3383;50° (= 9 · 360° + 143;50°), and 3383;50°/259d = 13;3,53,59°. Similarly, 287d · 13;3°/d is about 3745°. Hence the entry for 287d corresponds to 3748;56°, and 3748;56°/287 = 13;3,44,57°/d. The corresponding values in the Alfonsine Tables for these mean motions are: 13;10,35°/d and 13;3,53°/d. We cannot explain the discrepancies.

In this table we are also given the mean motion in longitude of the Moon for all integer hours from 1h (0;0,32°) to 24h (0;13,10°) and the mean motion in lunar anomaly for all integer hours from 1h (0;0,32°) to 24h (0;13,4°). An explanation is found at the bottom of the page: "Table for the days that the children will be in the womb of their mother. If they are born earlier or in the middle [*sic*], they will not live."

In MS Mu, f. 218a, there are two tables, Tables HG 70 and 71 (see Tables HG 70A and HG 71A).

On f. 218b there is an astrological table for the planetary 'terms', first according to the Egyptians, and then according to Ptolemy (cf. *Tetrabiblos*, I.20–21; ed. Robbins 1940, pp. 91 ff). Folio 219a contains a table of planetary aspects, and f. 219b some tables of astrologically significant places. The rest of MS Mu is blank, and ends at f. 229b.

Table HG 70A. Heading: "Table of the dignities (*koḥot*: *lit.* powers) of the planets in the zodiacal signs, and this is the knowledge of the planetary houses and their exaltations (*nesi'utam*) and their joys (*simḥatam*), and what are the diurnal and nocturnal lords of the triplicities, and who are the lords of the faces". A similar table is found in Segovia, MS 110, f. 84v.

I	II	III	IV	V	VI
Signs	**Houses/ Exalt.**	**Joys (*sic*)**	**Diurnal Lords of the trip.**	**Nocturnal Lords of the trip.**	**Lords of the faces**
Dign.	**5/4**		**3**	**3**	**1**
Ari	Mars/Sun	Sun	Sun/Jup/Sat	Jup/Sun/Sat	Mars/Sun/Ven
Tau	Ven/Moon	Moon	Ven/Moon/Mars	Moon/Ven/Mars	Merc/Moon/Sat
Gem	Merc/AscN	—	Sat/Merc/Jup	Merc/Sat/Jup	Jup/Mars/Sun
Cnc	Moon/Jup	—	Ven/Mars/Moon	Mars/Ven/Moon	Ven/Merc/Moon
.					
.					
Psc	Jup/Ven	Jup	Ven/Mars/Moon	Mars/Ven/Moon	Sat/Jup/Mars

1. For the dignities of the planets, see al-Bīrūnī, *Tafhīm* (Wright 1934, p. 306 f). The dignities are given in a scale of numbers: "5 to the house, 4 to exaltation, 3 to term, 2 to triplicity, and 1 to face". These dignities correspond, more or less, to the dignities in Table HG 70A.
2. For the planetary houses, see al-Bīrūnī, *Tafhīm* (Wright 1934, p. 256): Ari: Mars, Tau: Venus, . . . , Psc: Jupiter.
3. For the exaltations of the planets, see al-Bīrūnī, *Tafhīm* (Wright 1934, p. 258):

Saturn:	Lib	21
Jupiter:	Cnc	15
Mars:	Cap	28
Sun:	Ari	19
Venus:	Psc	27
Merc:	Vir	15
Moon:	Tau	3
Asc N:	Gem	3
Desc N:	Sgr	3

These values in al-Bīrūnī correspond to those in Segovia, MS 110, f. 84v, col. III. In MS Mu the degrees within each sign have been omitted.

4. The column for 'joys' has planets associated with zodiacal signs rather than with astrological houses as expected (see notes to Table HG 71A, below). The entries in this column agree with the houses (for the Sun and the Moon), or with the exaltations (for the 5 planets). There follows a comparison of the 'joys' in Table HG 70A with the planetary houses according to al-Bīrūnī, *Tafhīm* (Wright 1934, p. 256), and the exaltations of the Sun and the Moon (as above):

	Mu, 218a Table HG 70A	**al-Bīrūnī**	
Sign	**Joy**	**House**	**Exalt.**
Ari:	Sun	Mars*	Sun
Tau:	Moon	Venus*	Moon
Vir:	Mercury	Mercury	
Lib:	Venus	Venus	
Sco:	Mars	Mars	
Aqu:	Saturn	Saturn	
Psc:	Jupiter	Jupiter	

* According to al-Bīrūnī, the house for the Sun is Cnc, and the house for the Moon is Leo. Maybe the author of Mu got confused, and chose one of his first two entries to produce the third entry in each row, i.e., the first two joys are the exaltations, and the next five are the houses.

5. For a list of diurnal and nocturnal lords of the triplicities, see al-Bīrūnī, *Tafhīm* (Wright 1934, p. 259):

		diurnal lords	**nocturnal lords**
1st trip:	Ari, Leo, Sgr:	Sun/Sat	Jup/Sat
2nd trip:	Tau, Vir, Cap:	Ven/Mars	Moon/Mars
3rd trip:	Gem, Lib, Aqu:	Sat/Jup	Merc/Jup
4th trip:	Cnc, Sco, Psc:	Ven/Moon	Mars/Moon

These agree with the entries in Table 70A, cols. IV and V, but MS Mu has added a planet in the middle of each pair given by al-Bīrūnī.

6. For the lords of the faces, see al-Bīrūnī, *Tafhīm* (Wright 1934, p. 263):

	Lords of the faces
Ari:	Mars/Sun/Ven
Tau:	Merc/Moon/Sat
.	
.	
Psc:	Sat/Jup/Mars

Table HG 71A. Heading: "Table of the houses of the planetary joys (*simḥatam*), and the years that are counted for the natives (*ha-noladim*)". Segovia, MS 110, f. 82v, has a comparable list, with the heading: "Tabula conbinationis(?) nati. in regimine generali(?)", and adds a second row with the numbers 4, 14, 22, 41, 56, 68, and 78.

Planet	Moon	Merc	Ven	Sun	Mars	Jup	Sat
Houses	3rd House	1st	5th	9th	8th	11th	12th
Years that are counted [for natives]	4	10	8	19	15	12	old age (*ha-ziqna*)

1. For the joys of the planets, see al-Bīrūnī, *Tafhīm* (Wright 1934, p. 277, paragraph 469). The joy of a planet is assigned to an astrological house (determined with respect to the local horizon at a given moment in time) as follows:

	Mu, 218a	**al-Bīrūnī**
Moon	3rd	3rd
Merc	1st	1st
Ven	5th	5th
Sun	9th	9th
Mars	8th*	6th
Jup	11th	11th
Sat	12th	12th

* So all but Mars agree. The Arabic term that Wright translates as 'joy' is *faraḥ*. (Unfortunately, Wright transliterated it as *fahr*, but this is not a word in the Arabic dictionary, whereas *faraḥ* is a common word meaning 'joy', and this is the word that appears in al-Bīrūnī's Arabic text published by Wright.) The joy of a planet is not usually associated with a zodiacal sign, as it was in Table HG 70A.

2. For the number of years associated with the planets in Table HG 71A, see al-Bīrūnī, *Tafhīm* (Wright 1934, p. 255). Note that there is only partial agreement with Table HG 71A:

Planet	**Least years**
Sat	30
Jup	12
Mars	15
Sun	19
Venus	8
Merc	20
Moon	25

4. The *Almanach Perpetuum*

4.1 The dedication

The Latin version of the *Almanach Perpetuum* (1496) begins with a dedication, missing in the Castilian version, to an unnamed dignitary of the Church in Salamanca. Although it has not previously been noted, this dedication is preserved in two slightly different variants. In some copies the title is "Epistula actoris ad presbiterem salamantice," e.g., Madrid, Biblioteca Nacional, shelfmark I-1077 (formerly I-1350); whereas in others it is "Epistola actoris ad episcupum salamantice," e.g., Biblioteca Universidad de Salamanca, shelfmarks Inc. 176 and Inc. 177. The text in the second variant corrects some of the numerous spelling errors found in the first variant (see below). Cantera (1931, pp. 74–75) partially transcribed the dedication in the second variant, as it is found in the copies preserved in Salamanca, and gave a partial translation of it in Spanish (Cantera 1935, pp. 21–22). In the 1502 edition of the *Almanach Perpetuum* (Venice) the title is given as "Epistola auctoris ad episcopum Salamantice."

There follows a transcription of the dedication in the copy at the Biblioteca Nacional (Madrid), i.e., the first variant of the Latin edition of the *Almanach Perpetuum* (1496). We have compared it with another dedication, included in Regiomontanus's *Tabulae Directionum*, composed in Esztergom (1467) and first printed in Augsburg (1490), to the Hungarian Archbishop János Vitéz (d. 1472). Its title is: "Reuerendissimo in Christo patri et domino: domino Joanni archiepiscopo Strigoniense legato etc. Joannes Germanus de regiomonte se humiliter commendat." It is clear that the dedication in the *Almanach Perpetuum* reproduces, almost verbatim, part of Regiomontanus's text (to which we assign the siglum *R*), a fact already noticed by Cohn (1917, p. 106) and then by Zinner (1990, p. 121). In addition to changing the title, there are three passages in Regiomontanus's dedication that were not reproduced in the *Almanach Perpetuum* and, significantly, "Vienna" was replaced by "Salamanca" (see note 32, below). Moreover, the expression "vt tabulas quasdam directionum" in *R* has been replaced by "vt tabulas quasdam de locis planetarum" in the *Almanach Perpetuum* (see note 37, below), reflecting the difference in the content of the two works. The rest of the changes are minor and have to do with spelling and Latin grammar. Note that the quality of the Latin used in the dedication of the *Almanach Perpetuum* is very deficient in contrast to that of Regiomontanus.

The dedication printed in the 1502 edition of the *Almanach Perpetuum* follows that of 1496, but eliminates most of its typographical and spelling

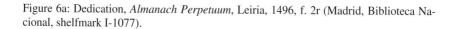

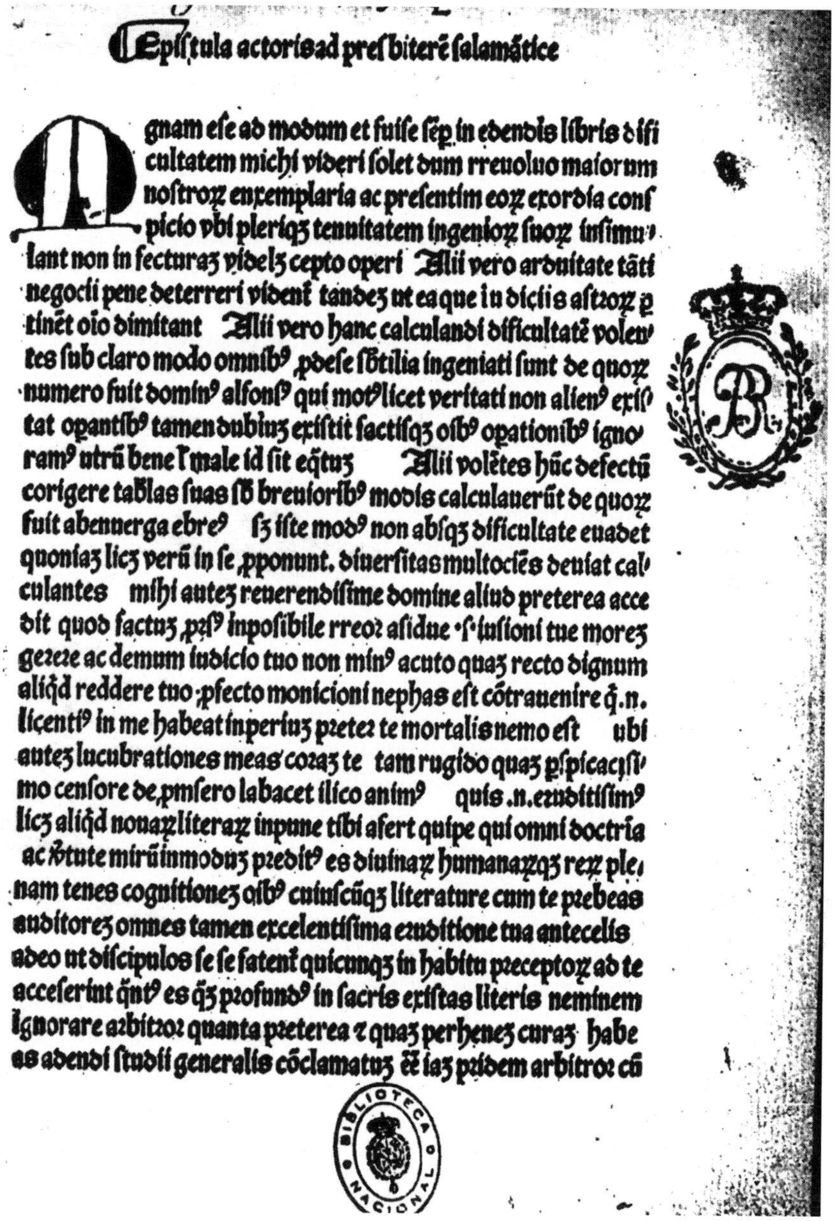

Epiſtula actoris ad preſbiterē ſalamātice

Magnam eſſe ad modum et fuiſſe ſēp in edendis libris difi
cultatem michi videri ſolet dum rreuoluo maiorum
noſtroꝝ exemplaria ac preſentim eoꝝ exordia conſ
picio vbi pleriqꝫ tenuitatem ingenioꝝ ſuoꝝ inſimu
lant non in fecturaꝫ videlꝫ cepto operi Alii vero arduitate tāti
negocii pene deterreri vident tandeꝫ ut ea que in diciis aſtroꝝ ꝑ
tinēt oio dimitant Alii vero hanc calculandi dificultatē volen
tes ſub claro modo omnibꝰ ꝓdeſſe ſbtilia ingeniati ſunt de quoꝝ
numero fuit dominꝰ alfonſꝰ qui motꝰ licet veritati non alienꝰ exiſ
tat opantibꝰ tamen dubiuꝫ exiſtit factiſqꝫ oibꝰ opationibꝰ igno
ramꝰ utrū bene ꝉ male id ſit eqtuꝫ Alii volētes hūc defectū
corigere tablas ſuas ſb breuioribꝰ modis calculauerūt de quoꝝ
fuit abenuerga ebreꝰ ſꝫ iſte modꝰ non abſqꝫ dificultate euadet
quoniaꝫ licꝫ verū in ſe ꝓponunt. diuerſitas multociēs deuiat cal
culantes mihi auteꝫ reuerendiſſime domine aliud preterea acce
dit quod factuꝫ ꝓpꝰ inpoſibile rreoꝛ aſidue ·ſ· iuſioni tue moreꝫ
gerere ac demum iudicio tuo non minꝰ acuto quaꝫ recto dignum
aliq̄d reddere tuo ꝑfecto monicioni nephas eſt cōtrauenire q̄.n.
licentiꝰ in me habeat inperiuꝫ preter te mortalis nemo eſt ubi
auteꝫ lucubrationes meas coraꝫ te tam rugido quaꝫ ꝑſpicaciſſi
mo cenſore deꝓmſero labacet ilico animꝰ quis .n. eruditiſſimꝰ
licꝫ aliq̄d nouaꝝ literaꝝ inpune tibi afert quipe qui omni doctria
ac ꝟtute mirū inmoduꝫ preditꝰ es diuinaꝝ humanaꝝqꝫ reꝝ ple
nam tenes cognitioneꝫ oibꝰ cuiuſcūqꝫ literature cum te prebeas
auditoreꝫ omnes tamen excelentiſſima eruditione tua antecelis
adeo ut diſcipulos ſe ſe fatenꝫ quicunqꝫ in habitu preceptoꝝ ad te
acceſerint q̄ntꝰ es q̄ꝫ profundꝰ in ſacris exiſtas literis neminem
ignorare arbitroꝛ quanta preterea ꝛ quaꝫ perheneꝫ curaꝫ habe
as adendi ſtudii generalis cōclamatuꝫ ē iaꝫ pridem arbitroꝛ cū

Figure 6a: Dedication, *Almanach Perpetuum*, Leiria, 1496, f. 2r (Madrid, Biblioteca Nacional, shelfmark I-1077).

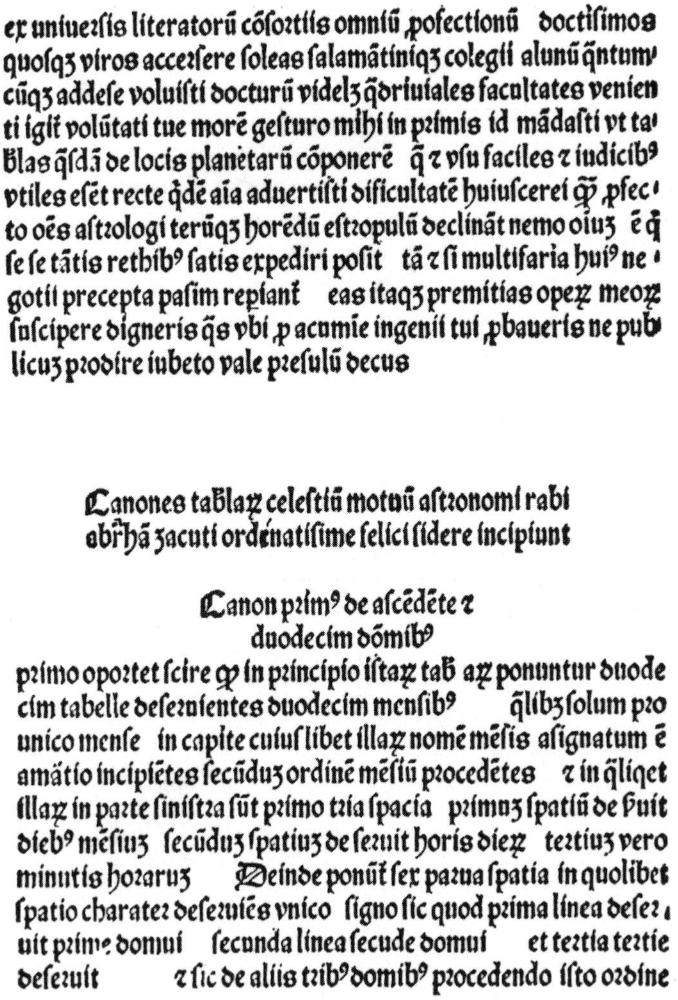

ex uniuerſis literatorũ cõſortiis omniũ ꝓfectionũ doctiſimos
quoſq3 viros accerſere ſoleas ſalamãtiniq3 colegii alunũ q̃ntum
cũq3 addeſe voluiſti docturũ videl3 q̃driuiales facultates venien
ti igit volũtati tue morẽ geſturo mihi in primis id mãdaſti vt ta
blas q̃ſdã de locis planetarũ cõponerẽ q̃ ⁊ vſu faciles ⁊ iudicib9
vtiles eſẽt recte q̃dẽ aĩa aduertiſti dificultatẽ huiuſcerei q̃ ꝑfec
to oẽs aſtrologi terũq3 horẽdũ eſtropulũ declinãt nemo oiu3 ẽ q̃
ſe ſe tãtis retbib9 ſatis expediri poſit tã ⁊ ſi multiſaria hui9 ne
gotii precepta paſim repiant eas itaq3 premitias opeꝝ meoꝝ
ſuſcipere digneris q̃s vbi ꝑ acumie ingenii tui ꝓbaueris ne pub
licu3 prodire iubeto vale preſulũ decus

Canones tablaꝝ celeſtiũ motuũ aſtronomi rabi
abrhã 3acuti ordenatiſime felici ſidere incipiunt

Canon prim9 de aſcẽdẽte ⁊
duodecim dõmib9

primo oportet ſcire q̃ in principio iſtaꝝ tab aꝝ ponuntur duode
cim tabelle deſeruientes duodecim menſib9 q̃lib3 ſolum pro
unico menſe in capite cuiuſ libet illaꝝ nomẽ mẽſis aſignatum ẽ
amãtio incipiẽtes ſecũdu3 ordinẽ mẽſiũ procedẽtes ⁊ in q̃liqet
illaꝝ in parte ſiniſtra ſũt primo tria ſpacia primu3 ſpatiũ de ſuit
dieb9 mẽſiu3 ſecũdu3 ſpatiu3 de ſeruit horis dieꝝ tertiu3 vero
minutis horaru3 Deinde ponũt ſex parua ſpatia in quolibet
ſpatio charater deſeruiẽs vnico ſigno ſic quod prima linea deſer
uit prime domui ſecunda linea ſecude domui et tertia tertie
deſeruit ⁊ ſic de aliis trib9 domib9 procedendo iſto ordine

Figure 6b: Dedication, *Almanach Perpetuum*, Leiria, 1496, f. 2v (Madrid, Biblioteca Nacional, shelfmark I-1077).

mistakes. In fact, it looks so close to Regiomontanus's text that one might easily imagine that the Venetian printers of the dedication in the 1502 *Almanach Perpetuum* copied it directly from Regiomontanus's *Tabulae Directionum*.

We have indicated the original passages in the *Almanach Perpetuum* by enclosing them in angle brackets < . . . >. In the notes, variant reading are given from Regiomontanus's *Tabulae Directionum*, ed. 1490; we have also consulted the ed. of 1504. We have not noted variant readings in punctuation which, in any event, is rarely used in the *Almanach Perpetuum*.

Base text: Madrid, Biblioteca Nacional, I-1077

Comparison with Regiomontanus's *Tabulae Directionum*, Augsburg (1490), here given the *siglum R*

Epistula actoris ad presbiterem salamantice

Magnam ese ad modum et fuise[1] semper in edendis libris dificultatem michi[2] videri solet dum rreuoluo[3] maiorum nostrorum enxemplaria[4] ac presentim[5] eorum exordia conspicio vbi plerique tenuitatem ingeniorum suorum insimulant non in fecturam[6] videlicet cepto operi Alii vero arduitate tanti[7] negocii pene deterreri vident[8] <tandem ut ea que iudiciis astrorum pertinent omnino dimitant Alii vero hanc calculandi dificultatem volentes sub claro modo omnibus prodese subtilia ingeniati sunt de quorum numero fuit dominus alfonsus qui motuslicet veritati non alienus existat operantibus tamen dubium existit factisque omnibus operationibus ignoramus utrum bene vel male id sit equatum Alii volentes hunc defectum corigere tabulas suas sub breuioribus modis calculauerunt de quorum fuit abenuerga ebreus Sed iste modus non absque dificultate euadet quoniam licet verum in se proponunt. diuersitas multociens deuiat calculantes>[9] mihi autem reuerendissime domine aliud preterea accedit quod factum[10] prorsus imposibile rreor asidue[11] scilicet iusioni[12] tue morem gerere ac demum iudicio tuo non

[1] esse admodum et fuisse *R*
[2] difficultatem mihi *R*
[3] reuoluo *R*
[4] exemplaria *R*
[5] presertim *R*
[6] suffecturam *R*
[7] tentati *R*
[8] videntur *R*
[9] nonnulli erratis suis veniam datum iri volunt dubiam scribendi fortunam haud iniuria suspicantes *R, instead of the passage placed within angle brackets*
[10] factu *R*
[11] impossibile reor assidue *R*
[12] iussioni *R*

minus acuto quam recto dignum aliquid reddere tuo[13] profecto monicioni[14] nephas est contrauenire q. n.[15] licentius in me habeat inperium[16] preter te mortalis nemo est ubi autem lucubrationes meas coram te tam rugido[17] quam perspicacisimo[18] censore de promsero labacet ilico animus[19] quis .n. eruditisimus[20] licet aliquid nouarum literarum inpune[21] tibi afert quipe[22] qui omni doctrina ac virtute mirum in modum preditus es diuinarum humanarumque rerum plenam tenes cognitionem omnibus cuiuscumque literature cum te prebeas auditorem omnes tamen excelentisima[23] eruditione tua antecelis[24] adeo ut discipulos se se fatentur[25] quicunque in habitu preceptorum ad te acceserint[26] quantus es quam profundus in sacris existas literis neminem ignorare arbitror[27] quanta preterea et quam perhenem[28] curam habeas adendi[29] studii generalis conclamatum esse iam pridem arbitror cum ex uniuersis literatorum consortiis omnium profectionum doctisimos[30] quosque viros accersere soleas[31] salamantinique colegii alunum[32] quantumcumque addese[33] voluisti docturum videlicet quadriuiales facultades[34] venienti igitur voluntati[35] tue morem gesturo mihi in primis id mandasti[36] vt tabulas quasdam <de locis planetarum>[37] componerem que et vsu faciles et iudicibus vtiles esent[38] recte quidem anima aduertisti dificultatem[39] huiuscerei quam profecto omnes astrologi terumque horendum estropulum[40]

[13] reddere. Tue *R*
[14] monitioni *R*
[15] qui enim *R*
[16] imperium *R*
[17] rigido *R*
[18] perspicacissimo *R*
[19] labascet illico annus *R*
[20] enim eruditissimus *R*
[21] impune *R*
[22] afferet quippe *R*
[23] excellentissima *R*
[24] antecellis *R*
[25] sese fateantur *R*
[26] accesserint *R*
[27] *A passage of about 9 lines in R is omitted here.*
[28] perhennem *R*
[29] condendi *R*
[30] professionum doctissimos *R*
[31] *In R there are 10 additional words:* officio fretus regii cancellarii supremi: cui cepto felicissimo: me quoque
[32] Wiennensis collegii alumnum *R*
[33] adesse *R*
[34] quadruuiales facultates *R*
[35] voluntatique *R*
[36] mandati dedisti *R*
[37] directionum *R, instead of the words within angle brackets*
[38] essent *R*
[39] animaduertisti difficultatem *R*
[40] tamque horrendum scopulum *R*

declinant nemo omnium est qui se se[41] tantis rethibus satis expediri posit[42] tam etsi multifaria huius negotii[43] precepta pasim[44] reperiantur[45] eas itaque premitias[46] operum meorum suscipere digneris quas vbi pro acumine ingenii tui probaueris ne[47] publicum prodire iubeto vale presulum decus.

Zinner (1990, p. 121) asserts that "Josef Vizinus translated the almanac from Hebrew to Latin in about 1484 and dedicated it to the Archbishop of Salamanca," but there is no evidence for this dating. Furthermore, there is no reason to believe that Zacut was aware of this dedication, or that there ever existed a Hebrew version of it. According to Cantera, this dedication was addressed to Gonzalo de Vivero, Bishop of Salamanca, presumed to be the patron of Zacut for several years. This bishop died in 1480, and it seems odd to dedicate a work to someone who had died 16 years earlier. Moreover, the text ends with "vale" which is only appropiate for addressing the living. In the dedication by Regiomontanus "vale" was indeed appropriate, for the archbishop to whom it was addressed was alive at the time he wrote it. This text has been the sole support for the claim that Zacut was either a teacher or a student at the University of Salamanca. But we can now see that the mention of the University of Salamanca here is simply inserted on analogy with Regiomontanus's remark concerning the University of Vienna, and tells us nothing about Zacut. There is also no reason to believe that Zacut was aware of the edition of Regiomontanus's *Tabulae Directionum* at any time, and certainly not in 1478 when his almanac was composed, 12 years prior to the *editio princeps* of Regiomontanus's work.

This dedication has inspired quite a bit of speculation concerning the life of Zacut, but none of it can be justified. We conclude that this dedication had nothing to do with the bishop of Salamanca, nor with Zacut, and was simply added by Vizinus, the "translator" according to the colophon, or by d'Ortas, the printer of the *Almanach Perpetuum*, to make it conform with other scientific works that had recently been published.

4.2 The canons

Two versions of the *Almanach Perpetuum* were produced by the printing house of Samuel d'Ortas[48] in 1496: one has the canons in Castilian, the

---

[41] sese *R*
[42] expedire possit *R*
[43] negocii *R*
[44] passim *R*
[45] *A passage of 38 lines in R is omitted here.*
[46] primicias *R*
[47] in *R*
[48] Roth 1954. For the Hebrew works printed by d'Ortas, see Aron Freimann 1925.

other in Latin.[49] At the end of the Castilian canons we are told that Vizinus translated the text from Hebrew into Latin, and then from Latin into *noestro vulgar romançe*, meaning Castilian, even though Vizinus is supposed to be Portuguese and the book was printed in Portugal. Moreover, the *Almanach Perpetuum* is the first scientific treatise to be printed in Portugal and, oddly enough, the text is in Castilian rather than Portuguese. On the other hand, the canons are not a translation from Hebrew, as will be argued below.

The Castilian version is arranged in 23 chapters (30 pages of 30 lines each), and the Latin version in 12 chapters. There were two printings of the canons for the Latin version: in one printing they are presented in 20 pages of 32 lines each, and in the other in 27 pages of 29 lines each.

There is also a manuscript copy of the Castilian version preserved in Seville, Biblioteca Colombina, MS 5-2-32, ff. 48r-77v and, according to Cantera (1931, p. 135), it was written in the fifteenth century. The manuscript contains the canons but no tables, and follows the printed edition quite closely, although there are at least 20 places where they differ (sometimes there are expansions of explanations, and sometimes the manuscript omits a few words). Of particular interest is that the explicit in the manuscript (f. 77v) is shorter than in the edition: in the following passage, the words in parentheses are not found in the manuscript.

> Aqui se acaba la reçela delas tablas tresladadas de abrayco en latin en (noestro vulgar) romance por maestre Jusepe (vezino deçipolo del actor de las tablas) Deo graçias.

It is likely that this manuscript was copied from the printed edition.

Next, let us compare the Latin and Castilian versions for the canons of the *Almanach Perpetuum*, and their relationship with the canons of the *Ḥibbur*. The chapters in the Castilian version have the following headings:

Capitulo primero para saber el asçendente e las doze casas
Capitulo 2 para saber el lugar del sol
Capitulo terçero para saber la entrada del sol en qual quiera delos signos
Capitulo quarto para saber el verdadero lugar dela luna
Capitulo quinto para ygualar la luna para cada hora
Capitulo sexto para ygualar la cabeça del drgon (*sic*)
Capitulo septimo para saber el movimiento de latitud dela luna

[49] For the Castilian version we have used the facsimile reproduction of the copy at the Biblioteca Nacional (Lisbon), in Albuquerque 1986. In Bensaude 1919, pp. 1–35, there is a facsimile reproduction of the Castilian canons in the copy at the Evora Library, and another in Flórez *et al.* 1989, pp. 135–168, which is a facsimile of the copy at the Biblioteca Colombina (Seville). For the Latin version we have used the copy at the Biblioteca Nacional (Madrid), shelfmark I-1077 (formerly I-1350); see Cantera 1935, p. 58 ff. For a list of the editions and extant manuscripts, see Cantera 1935, p. 57 ff., and Fontoura 1983, pp. 426 ff.

Capitulo seteno para saber la coniunçion e oposiçion del sol e dela luna
Capitulo de coniunçiones e oposiçiones tiradas por las tablas dela luna
Capitulo 8 delos eclipses e primero delos del sol
Capitulo delos eclipses dela luna
Capitulo 9 para igualar el verdadero lugar de saturno
Capitulo 10 para saber el movimiento de saturno para cada día
Capitulo para saber el movimiento dela latitud de saturno
Capitulu (*sic*) para ygualar el çentro e el argumento despues de pasada la primera revoluçion
Capitulo para saber el verdadero movimiento de jupiter
Capitulo 14 para saber el movimiento dela latitud de jupiter
Capitulo 15 para saber el verdadero lugar de jupiter
Capitulo 16 para saber el verdadero lugar de martis
Capitulo 17 para saber el movimiento de maris (*sic*) en cada un día
Capitulo 18 para saber el movimiento dela ladeza de mars
Capitulo para saber el logar (*sic*) verdadero de venus
Capitulo 20 para saber el verdadero muvimiento (*sic*) de latitud de venos (*sic*)
Capitulo 21 del verdadero lugar de mercurio
Capitulo 22 para saber el movimiento dela ladeza de mercurio
Capitulo 23 de animodar

The Latin version of the same text is divided differently into canons as follows:

Epistula actoris ad presbiterem salamantice
Canon primus de ascendente et duodecim domibus
Canon secundus de vero loco solis habendo
Canon tertius de introitu solis in quolibet signorum
Canon quartus de loco vero lune habendo
Canon quintus de coniuntionibus et opositionibus luminarium
Canon sextus de eclipsibus luminarium et primo de sole
Canon septimus de vero loco saturni
Canon sextus (*sic*) de vero loco Jovis per has tabulas habendo
Canon nonus de vero loco martis habendo
Canon decimus de vero loco veneris habendo
Canon undecimus de vero loco mercurii per has tabulas invenire
Canon ultimus de animodar

Despite this apparent difference, the order of the subjects treated in both versions is the same, for the Latin has unnumbered *regulae* within the canons that deal with the remaining matters. The two versions are so close that there can be no doubt that one has been translated from the other. However, when we compare the canons of the *Almanach Perpetuum*, whether in Castilian or in Latin, with those in the *Ḥibbur* the differences are very striking, to such an extent that one cannot be considered a translation of the other. These differences are pervasive; some examples will suffice.

Only a few predecessors are cited in the *Almanach Perpetuum*, whereas the *Ḥibbur* mentions a much greater number of previous authors throughout the text, as indicated in chap. 2.6.

The *Almanach Perpetuum* has a list of movable feast days and a dedication to a bishop, both taken from Regiomontanus and lacking in the *Ḥibbur.* On the other hand, the introduction to the *Ḥibbur* gives references to the Talmud, and to the application of astronomy in Jewish law. Moreover, chapter 18 describes the principles of the Jewish calendar. It is surely significant that there are 19 chapters in the *Ḥibbur* and the Jewish calendar is based on a cycle of 19 years. These distinctions strongly suggest that the *Almanach Perpetuum* was intended for a Christian audience, whereas the *Ḥibbur* for a Jewish readership.

The *Ḥibbur* presents a much higher level of detail than the *Almanach Perpetuum.* So, for example, in chapter 9 of the *Ḥibbur* Zacut mentions an occultation of Spica by the Moon that he observed in Salamanca in 1474, from which he derived a value for precession. None of this is found in the *Almanach Perpetuum*, and many more examples of this kind could be cited.

In the introduction to the *Ḥibbur* Zacut gives a number of days in base 30 notation, and in chapter 18 this notation is explained. A few tables use this notation as well (see, e.g., Table HG 63). None of this is found in the *Almanach Perpetuum.*

The final chapter in the *Almanach Perpetuum* concerns an astrological procedure for nativities called *animodar*, and it has no counterpart in the *Ḥibbur*.

In sum, the *Ḥibbur* and the *Almanach Perpetuum* are distinct works, and we see no persuasive evidence that Zacut was involved in the edition of the latter.

4.3 The tables

The tables in the *Almanach Perpetuum* are listed below together with the page or folio numbers in the two versions of 1496. For the Castilian version we have consulted the copy in Lisbon, Biblioteca Nacional, and for the Latin version the copy in Madrid, Biblioteca Nacional. These tables take up more than 300 pages in both versions. Note that the Latin version has 3 extra tables on 2 folios, not included in the Castilian version.

The tables are the same in both the Latin and the Castilian versions (except as noted below), and all headings are in Latin. Among the copies

	Castilian Version	Latin Version
1. Ascendants and houses	109-114	18r-20v
2. Daily solar positions	115-122	21r-24v
3. Solar declination	123	25r
4. Solar correction	123	25r
5. Equation of time	124	25v
6. Entry into the zodiacal signs	125-130	26r-29v
7. Daily lunar positions	131-192	30r-60v
8. Syzygies	193-208	155r-162v
9. Mean motion in lunar anomaly	209-210	61r-v
10. Equation of center	210	61v
11. Nodes	211-212	62r-v
12. Solar elongation from the lunar node	213-214	63r-v
13. Equation of eclipse	213	63r
14. Parallax	215-216	64r-v
15. Solar eclipses	217	65r
16. Lunar eclipses	217	65r
17. Equation of syzygies	218-220	65v-66v
18. Lunar latitude	221, 223	67r-v
19. Right ascension	222	68r
20. Longitude of Saturn	224-229	68v-71r
21. Center of Saturn	230-235	71v-74r
22. Anomaly of Saturn	236-241	74v-77r
23. Latitude of Saturn	242-245	77v-79r
24. Longitude of Jupiter	246-263	79v-85r, 86r-88v
25. Center of Jupiter	264-272	85v, 89r-92v
26. Anomaly of Jupiter	272-284	92v-98v
27. Latitude of Jupiter	285-288	99r-100v
28. Longitude of Mars	289-304	101r-108v
29. Correction of Mars	305	109r
30. Center of Mars	306-310	109v-111v
31. Anomaly of Mars	311-315	112r-114r
32. Latitude of Mars	316-319	114v-116r
33. Longitude of Venus	320-327	116v-120r
34. Correction of Venus	328	120v
35. Center of Venus	328	120v
36. Anomaly of Venus	329	121r
37. Latitude of Venus	330-335	121v-124r
38. Longitude of Mercury	336-373	124v-143r
39. Correction of Mercury	374	143v
40. Center of Mercury	374	143v
41. Anomaly of Mercury	375-380	144r-146v
42. Unequal motion of Mercury	381-386	149r-151v
43. Latitude of Mercury	387-396	147r-148v, 152r-154v

Continued

	Castilian Version	Latin Version
44. Sexagesimal multiplication	397-399	163r-164r
45. Fixed stars	400-401	164v-165r
46. Animodar	402-404	165v-166v
47. Solar and lunar eclipses	—	167r
48. Half-length of daylight	—	167v
49. Geographical coordinates	—	168r-v
50. Calendar	405–408	169r-170v
51. Sunday letters	409	171r
52. Movable feast days	410–411	171v-172r

of the Latin version we have consulted, there are some differences in the ordering of the tables,[50] thus suggesting that the Latin version was printed in three separate runs. Moreover, some minor corrections and changes in the sequence of the tables suggest that the Castilian version was printed before the Latin. There were several subsequent editions of the *Almanach Perpetuum* (see chap. 5.2).

The tables in the *Almanach Perpetuum* are here preceded by the *siglum* "AP."

AP 1. Ascendants and houses

The use of this table is described in chapter 8 of the *Ḥibbur*, and in chapter 1 of the Leiria edition. It contains 12 monthly sub-tables, beginning in March. For each day of the year we are given (i) true solar time (in hours and minutes); (ii) the longitude, at noon at Salamanca, of the cusps of the first 6 astrological houses (in signs and degrees). By combining both sets of data, the table can be used to determine the longitudes of the cusps at any time by means of the following rule: if $H(x)$ is the hour angle of the Sun for a given date x, and λ_1 is an entry for the longitude of the ascendant at noon, then $\lambda_1(H(x) \pm t)$ gives the ascendant at a time t after (+), or before (−), noon of day x.

Both the Castilian and the Latin versions give examples for the use of that table: the determination of the ascendant at noon and at 2 p.m.

[50] In the copy of the Latin version in Salamanca, Biblioteca Universitaria (Inc. 177), tables 8, 9, 10, and 11 follow the same sequence as in the Castilian version, whereas in another Latin copy in the same library (Inc. 176) the sequence is 9, 10, 11, and 8; thus, the ordering in these two copies differs from that in the copy in Madrid, Biblioteca Nacional. Moreover, in contrast to the copy in Madrid, table 45 in the two copies in Salamanca is divided into two parts, and between them we find tables 49, 47, and 48 (in that order) in Inc. 177, and tables 47, 48, and 49 in Inc. 176. The tables for the longitude and center of Jupiter (tables 24 and 25) are correctly bound in the Salamanca copies. One last obvious difference: just after the colophon, Inc. 176 exhibits a round black seal with a coat of arms and a name, "IODE VIZINHO"; the seal is missing in Inc. 177.

(Castilian version), but at noon and at 4 p.m. (Latin version), for the same day, December 15. It should be noted that the example in chapter 8 of the *Ḥibbur* concerns the ascendant at 3;20 p.m. at Salamanca on August 12, which is given as Sgr 28°. August 12 was already mentioned in the *Ḥibbur*, for it appears in the very first example given in that treatise (see chap. 1, above).

We have recomputed the entries of this table in different ways, and the best results are obtained when using the values in the text for the obliquity of the ecliptic and the latitude of the place for which the table was computed: $\varepsilon = 23;33°$, and $\phi = 41;19°$ (latitude of Salamanca). Among the various methods for domification used by medieval astronomers, the "standard" method gives the best results (see Table AP 1A).[51] We present below some results of our computations for the determination of the longitudes of the cusps of houses 2 to 6 (λ_2 to λ_6), for a given value of the ascendant (λ_1). Only the results for the beginning of each month are shown. In each case, the first line displays the entries in the text, whereas the following lines show computations for different values of λ_1 within an interval of $\pm$ 0;30° around the value of the ascendant given in the text.

In all but 5 cases (out of 60), we find perfect agreement with the entries in the text, if we allow the ascendant to vary within a range of $\pm$ 0;30° around the value for λ_1 given in the text. This suggests that the compiler of the table used more accurate values of λ_1 when computing the longitudes of the houses than those he displayed, and that the values in the final version of his table have been rounded.

AP 2. Table for the daily solar positions

This table displays the true longitudes of the Sun given to seconds, and calculated for the meridian of Salamanca for noon of each day of a four-year cycle beginning on March 1, 1473. The use of this table is described in chapter 2 of the *Almanach Perpetuum* in both the Castilian and the Latin versions, as well as in chapter 1 of the *Ḥibbur* (Cantera 1931, p. 157).

Table AP 2 is essentially the same as Table HG 1 for which there are two versions, as noted above. But this table is a "hybrid" version, for it agrees with both versions of Table HG 1 for the first year; the version represented by Mu and W (Hebrew) in the next year and a half (from March 1, 1474, to September 1, 1475); and the version represented by L (Hebrew), Ac, Ma, and Se (Latin) in the last part of the table (from September 1, 1475 to the end). Surprisingly, the edition of the *Almanach Perpetuum* in 1498 follows the version in Mu and W in its entirety, rather than the hybrid version in

[51] For this and other methods, see North 1986. We have used a modified version of the computer program given at the end of North's book kindly provided to us by the author, to whom we are most grateful.

Table AP 1A: The Cusps of the Astrological Houses

		λ_1	λ_2	λ_3	λ_4	λ_5	λ_6
Mar 1	Cnc 12 =	102	123	146	170	210	247
		102; 0	123;25	146;11	170;21	210;35	247;34
		101;30	122;53	145;35	169;42	209;59	247; 2
Apr 1	Leo 5 =	125	149	175	201	238	271
		125; 0	149;20	175;09	201;17	237;51	271;21
May 1	Leo 26 =	146	175	203	230	262	293
		146; 0	173;53	202;12	229; 7	261;18	292;35
		146;30	174;28	202;50	229;44	261;50	293; 5
Jun 1	Vir 21 =	171	202	232	259	288	318
		171; 0	202;23	232; 4	259;31	288; 7	318;14
		170;30	201;50	231;30	258;56	287;34	317;42
Jul 1	Lib 15 =	195	229	259	288	315	345
		195; 0	228; 2	258;19	287;34	314;48	344;18
		195;30	228;33	258;52	288; 9	315;23	344;51
Aug 1	Sco 9 =	219	253	285	318	345	13
		219; 0	252;28	284;10	317;19	344; 8	11;58
		219; 0	252;58	284;43	317;58	344;46	12;33
Sep 1	Sgr 2 =	242	276	310	348	14	39
		242; 0	275;49	310;19	348;17	13;54	38;45
Oct 1	Sgr 24 =	264	299	337	17	41	63
		264; 0	298;46	336;37	17;11	40;49	62;59
Nov 1	Cap 23 =	293	329	9	48	70	92
		293; 0	329;30	9;10	47;41	70; 4	91;30
Dec 1	Psc 9 =	339	15	48	79	105	131
		339; 0	14;11	47;58	78;54	104;15	130;34
		339;30	14;38	48;19	79;10	104;35	130;59
Jan 1	Tau 8 =	38	64	87	111	144	181
		38; 0	63;22	87;16	111; 0	144;42	181;26
		37;30	62;57	86;55	110;41	144;18	180;58
Feb 1	Gem 17 =	77	98	121	142	182	221
		77; 0	97;53	119; 9	141;39	181;27	221; 3
		77;30	98;22	119;38	142; 9	182; 0	221;34

the *editio princeps* of 1496 that appears in all the other editions (see chap. 5.2, below).

Some historians have stressed the importance of this solar ephemeris during the period of the great discoveries, and they have considered it as the basis for the tables of daily solar declination used by pilots in the late fifteenth and early sixteenth centuries for navigation on the high seas (see, e.g., Albuquerque 1988). It should be pointed out, however, that the pattern used by Zacut for his 4-year table of the daily positions of the Sun was already common in medieval western Europe, and it is found in quite a few almanacs: Azarquiel's (epoch: September 1, 1088), Jacob ben Makhir's (March 1, 1301), Almanac of 1307 (March 1, 1307), to mention just a few.[52] All these almanacs, except for Azarquiel's, give the daily solar longitude to seconds, as in Zacut's table. Other tables of this kind, beginning on March 1, are found in the Humeniz-based *Tabule solis* compiled in Paris in 1239 and in the treatise on the *quadrans vetus* by Robertus Anglicus for years 1292–1295; in both cases, the table for the solar position is associated with a table for the solar declination.[53] Even in the Alfonsine *Libros del saber de astronomía* there is a similar table, although it is restricted to only one year beginning in January (Rico y Sinobas 1863–1867, 2:291–292).

In the *Almanach Perpetuum*, the third year (beginning in March) of the four-year cycle is a leap year, whereas in the *Ḥibbur* the leap year is the fourth (beginning in January). The *Almanach Perpetuum* thus follows the pattern of Jacob ben Makhir's almanac. On the other hand, in the Almanac of 1307 the leap year is the first year of the cycle, and in the Almanac of Azarquiel the leap-day is added at the end of August of the second year. In the four-year cycle of Robertus Anglicus, the last year is a leap year.

As the recomputation below shows, the values tabulated in the *Almanach Perpetuum* derive from the Alfonsine Tables as presented in the *editio princeps* (Ratdolt 1483). Before performing the recalculation, one needs to take into consideration that the Alfonsine Tables were established for Toledo and that the difference in longitude between Toledo (28;30°) and Salamanca (25;46°) is 2;44°, which is equivalent to a time difference of 0;0,27,20d. The longitudes of these two cities are attested in the geographical tables in the manuscripts of the *Ḥibbur* and in the Latin version of the *Almanach Perpetuum* (f. 168r-v), but not in the Castilian version, where there is no table of geographical coordinates.

In the following example, based on the Alfonsine Tables (1483), all quantities are expressed in sexagesimal form and given in degrees, except for the date which is presented in days since the Incarnation. On the method for recomputing the entries, see Poulle 1984, p. 201.

The value for the longitude of the solar apogee computed here (1,30;55,36°) is confirmed by the Hebrew manuscripts of the *Ḥibbur*

---

[52] For the Almanac of Azarquiel, see Millás 1943–1950; for the Almanac of Jacob ben Makhir, see Boffito and Melzi d'Eril 1908 (note that Dante was the owner of the copy they used for the publication, which is why his name is in the title of their work); for the Almanac of 1307 and its derivatives, see Chabás 1996b.

[53] For the *Tabule solis*, see Knorr 1997, esp. pp. 190–191. As for the table associated with the *quadrans vetus*, among the various MSS containing this table, we have examined Salamanca, Biblioteca Universitaria, MS 2662, ff. 47v-49r, and Bruges, MS 522, ff. 57v-59r; see also Poulle 1969, pp. 10–11.

March 1, 1473 (noon)	2,29,21,48; 0,27,20
Mean mot. acc. & rec. 8th sphere	1,14;55,14,42
Equation mean motion acc. & rec.	+8;41,16
Mean motion of the apogee	10;48,57,15
Longitude of the apogee	1,30;55,36
Mean motion of the Sun	5,48;18,39,58
Solar anomaly	4,17;23, 4
Solar equation	+2;17,51
True longitude of the Sun	5,50;26,31
	[= 11s 20;26,31]
Entry in the *Almanach Perpetuum*	Psc 20;26,30
	= 11s 20;26,30]
Difference (Zacut - Recomputation)	−0; 0, 1

(see comments to Table HG 51, where we note that the three Hebrew manuscripts and MS Ac all give the solar apogee as 3s 0;56°).

We have recomputed all the entries for March 1473 and obtained very good agreement, for the discrepancies between the entries in Zacut's table and our recomputation (Z − C) amount only to a few seconds in the worst case. The distribution of our results follows.

Z − C	$\leq -3''$	$-2''$	$-1''$	$0''$	$+1''$
Frequency	6	3	7	14	1

We have no explanation for the asymmetric distribution of these discrepancies.

AP 3. Table for the solar declination

This table gives the solar declination (δ) to minutes for each integer degree of solar longitude (λ). These magnitudes are related by the following formula, where ε is the obliquity of the ecliptic:

$$\delta = \arcsin(\sin\lambda \cdot \sin\varepsilon). \qquad [1]$$

The maximum entry is 23;33°, and corresponds to a value for the obliquity of the ecliptic which is associated with the *Mumtaḥan* zij of the ninth century (Vernet 1956). This kind of table, with the same maximum value, is quite common in the astronomical literature of medieval Spain. Indeed, it is mentioned in chapter 1 of the *Ḥibbur*, where we are told that it follows the "opinion of Azarquiel" (Cantera 1931, p. 158), and it is the value we find in the table for the declination in the Almanac of Azarquiel (Millás

Tabula prima solis cuiꝰ radix ē anno 1473

dies mensis	martiꝰ pisces g	m	z	aplis aries g	m	z	maius taurus g	m	z	iunius gemini g	m	z	iulius cancer g	m	z	augustꝰ leo g	m	z
1	20	26	30	20	54	0	19	51	7	19	25	4	17	55	52	17	32	38
2	21	25	59	21	52	24	20	48	36	20	22	5	18	52	55	18	30	17
3	22	25	28	22	50	48	21	46	5	21	19	7	19	49	58	19	27	56
4	23	24	56	23	49	8	22	43	34	22	16	8	20	47	2	20	25	36
5	24	24	21	24	47	28	23	41	2	23	13	9	21	44	6	21	23	18
6	25	23	46	25	45	48	24	38	30	24	10	11	22	41	12	22	21	0
7	26	23	11	26	44	0	25	35	54	25	7	12	23	38	19	23	18	42
8	27	22	26	27	42	11	26	33	17	26	4	13	24	35	26	24	16	32
9	28	21	41	28	40	22	27	30	40	27	1	15	25	32	37	25	14	22
10	29	20	55	29	38	26	28	28	0	27	58	17	26	29	48	26	12	12
11	♈0	20	3	♉0	36	30	29	25	19	23	55	19	27	27	0	27	10	4
12	1	19	11	1	34	35	♊0	22	38	29	52	20	28	24	13	28	7	57
13	2	18	10	2	32	32	1	19	54	♋0	49	21	29	21	26	29	5	50
14	3	17	18	3	30	29	2	17	10	1	46	22	♌0	18	40	♍0	3	53
15	4	16	16	4	28	25	3	14	25	2	43	23	1	15	59	1	1	56
16	5	15	14	5	26	16	4	11	37	3	40	24	2	13	18	2	0	0
17	6	14	7	6	24	7	5	8	49	4	37	25	3	10	37	2	58	5
18	7	13	0	7	21	58	6	6	0	5	34	26	4	7	58	3	56	11
19	8	11	53	8	19	44	7	3	9	6	31	28	5	5	10	4	54	17
20	9	10	40	9	17	29	8	0	18	7	28	30	6	2	40	5	52	36
21	10	9	25	10	15	14	8	57	27	8	25	31	7	0	0	6	50	54
22	11	8	10	11	12	54	9	54	32	9	22	32	7	57	33	7	49	14
23	12	6	52	12	10	34	10	51	36	10	19	34	8	55	0	8	47	36
24	13	5	34	13	8	14	11	48	40	11	16	35	9	52	28	9	45	58
25	14	4	16	14	5	51	12	45	44	12	13	37	10	49	57	10	44	20
26	15	2	51	15	3	27	13	42	48	13	10	39	11	47	26	11	42	49
27	16	1	26	16	1	3	14	39	51	14	7	41	12	44	57	12	41	18
28	17	0	1	16	58	35	15	36	54	15	4	43	13	42	28	13	39	48
29	17	58	32	17	56	6	16	33	57	16	1	46	14	40	0	14	38	20
30	18	57	3	18	53	37	17	31	0	16	58	49	15	37	32	15	36	52
31	19	55	34	0	0	0	13	28	2	0	0	0	16	35	5	16	35	24

3

Figure 7. Table AP 2: Table for the daily solar positions, first page only, *Almanach Perpetuum*, Leiria, 1496, f. 21r (Madrid, Biblioteca Nacional, shelfmark I-1077).

1943–1950, p. 174). The same value is also found, for example, in Ibn al-Kammād's zij, and in the Tables of Barcelona[54]. Comparison with these tables shows that Zacut's table for the solar declination was not copied from any of them.

This table, when associated with that for the daily solar positions, is considered the basis for the rutters used by the Portuguese pilots for what has been called "astronomical navigation." The oldest rutters preserved are the *Regimento de Munich* (1509?) and the *Regimento de Évora* (1519), both using $\varepsilon = 23;33°$ and probably inspired by Zacut's work.[55]

We recomputed the entries in this table using equation [1], above, with $\varepsilon = 23;33°$, and we found the following distribution (Z − C is the difference between the entry in Zacut's table and our recomputation):

Z − C	−1′	0′	+1′
Frequency	2	76	12

AP 4. Table for the solar correction

In the *Almanach Perpetuum* the title of this table is *Tabula equationis solis*. This table corresponds to Table HG 2. Chapter 1 of the *Ḥibbur* describes the use of this table, valid for 34 four-year cycles (136 years). Chapter 2 of both the Castilian and the Latin versions mentions a correction of 0;1,46° to be added to the true position of the Sun after 4 years have elapsed.

From the entry for cycle 34 it is easy to deduce a more precise value for 1 cycle (4 years): 0;1,45,32,40°. This is exactly the length of the arc traveled by the Sun, at a constant speed of 0;59°/d, in 0;42,56h, which is the time difference between four Julian and four tropical years. As far as we know, the Portuguese scholar Antonio Barbosa (1928) was the first to explain this correction.

The length of the tropical year resulting from this correction of 1° for 136 years is 365;14,33,8,38d, in very good agreement with the underlying value in the *editio princeps* of the Alfonsine Tables (f. d3r): 365d 5;49,15,59,34,3h (= 365;14,33,9,58, . . . d).[56]

In Alfonso de Córdoba's edition of the *Almanach Perpetuum* (Venice 1502), the same table appears with some slight variations. All the entries are exact multiples of 0;1,46° (the tabulated value for cycle 34 is thus 1;0,4°,

[54] For Ibn al-Kammād's *al-Muqtabis,* see Chabás and Goldstein 1994. For the Tables of Barcelona, see Millás 1962 and Chabás 1996a.

[55] For transcriptions of these two rutters, see Bensaude 1912, p. 217 ff.

[56] For the Alfonsine value of the tropical year, see Swerdlow 1977; for the Alfonsine theory of precession with a fixed value for the tropical year and a variable value for the sidereal year, see Goldstein 1994.

Table AP 4A: The Solar Correction

4-year cycle	Correction in degrees	4-year cycle	Correction in degrees
1	0; 1,46	18	0;31,46
2	3,32	19	33,32
3	5,18	20	35,18
4	7, 4	21	37, 4
5	8,50	22	38,50
6	10,36	23	40,36
7	12,22	24	42,22
8	14, 8	25	44, 8
9	15,54	26	45,54
10	17,40	27	46,40*
11	19,25	28	49,25
12	21,11	29	51,11
13	22,57	30	52,57
14	24,43	31	54,43
15	26,59*	32	56,29
16	28,15	33	58,15
17	0;30, 0	34	1; 0, 0

* Note that the entries for cycles 15 and 27 were misprinted in both the Castilian and Latin versions: the entries should be 0;26,29° and 0;47,40°, respectively. Madrid, MS 3385, which contains a Latin version of the tables of the *Ḥibbur*, has the correct value for cycle 15, but shares the same error for cycle 27. In still another Latin version of the tables in the *Ḥibbur*, Segovia, MS 110, we find the correct values for cycles 15 and 27 (cf. Table HG 2).

rather than 1;0,0°, as in the edition of 1496), but the entry for cycle 35, 1;1,46°, is consistent with the value for cycle 34 in the 1496 edition rather than with the corresponding entry in the 1502 edition.

The Venice edition (1525) of the *Almanach Perpetuum* presents a table much like that in Alfonso de Córdoba's edition of 1502. There are entries for 35 cycles and all entries are exact multiples of 0;1,46°. According to Cantera, who depended on Barbosa, those who used Zacut's table for the solar correction in the sixteenth century included the astronomers Pedro Nunes, Gemma Frisius, and Johann Stoeffler (Cantera 1935, p. 122; on Nunes, see chap. 5.4.2).

The same correction of 1° for 136 years is found in a work written almost one century prior to Zacut, and it also depended on the Alfonsine Tables: Nycholas de Lynn's *Kalendarium* for 1387–1462. In chapter 1 the parameter is stated explicitly: *in 136 annis sol vix per unum gradum anticipatur* (Eisner 1980, p. 185:22–23), and the corresponding table, very similar to Zacut's, is entitled *Tabula continuacionis motus solis* (Eisner 1980, p. 180). Nycholas de Lynn's table gives the four-year corrections from 1385 to 1469, but the entries do not quite agree with those in Zacut's *Almanach Perpetuum* (e.g., 0;1,44° for cycle 1 [year 1389] rather than Zacut's 0;1,46°, and 0;34,51° for cycle 20 [year 1465] rather than Zacut's 0;35,18°).

AP 5. Table for the equation of time

The equation of time tabulated here is the difference between true and mean noon.[57] This table is identical with that in the *Ḥibbur* in Lyon, MS Heb. 14, f. 181r, and in Warsaw, ZIH, MS 245, f. 22b (Table HG 4). It is mentioned in chapter 1 of the *Ḥibbur*, where there is a short explanation for the conversion from true to mean solar time, and vice versa (Cantera 1931, p. 158). The entries are given in minutes of time for each day of the year, beginning in March. The extremal values of this function are:

max	= 0;22h:	27 April–21 May
min	= 0;12h:	12–25 July
Max	= 0;32h:	20–24 October
Min	= 0; 0h:	22 January–6 February.

These values are compatible with those in the corresponding table in the Alfonsine Tables (1483) which, in turn, is identical with that found in al-Battānī's zij and the Toledan Tables. All the entries in this table are positive (i.e., true noon always precedes mean noon), because the choice made for the value of the equation of time for the beginning of February is 0;0h here.

The format of this table is unusual in two ways. First, the argument is the day of the year, whereas in most medieval tables for the equation of time, it is the solar longitude (e.g., Ptolemy's *Handy Tables*, al-Battānī's zij). Second, the entries are not given in units of arc as in most other tables, but in units of time, a feature shared with the *Handy Tables*, where they are displayed to seconds. In Levi ben Gerson's table for the equation of time, the argument is the true longitude of the Sun and the entries are given in time degrees, where the maximum is 8;13° for Sco 9° and 10° (Goldstein 1974, pp. 102–104, 167). On the other hand, in the planetary tables of Immanuel ben Jacob Bonfils (*ca.* 1350) the equation of time is displayed in units of time (to minutes and seconds) where the argument is the longitude of the Sun, and the maximum is 0;33,20h for Sco 3° to 8° (Munich, MS Heb. 343, f. 182b, and MS Heb. 386, f. 15b; but in another copy of these tables the entries are in time degrees, where the maximum for the same arguments

[57] For a discussion on the equation of time in medieval astronomy, see Kennedy 1988, and van Dalen 1993, pp. 97–152. On the equation of time in Ptolemy, see Neugebauer 1975, pp. 61–68.

is 8;20°: see New York, JTSA, MS Heb. 2597, f. 59b). In chapter 5 of the *Ḥibbur*, Zacut mentions the discussion on the equation of time by Bonfils (see also Solon 1970, p. 10).

The *Tabule Verificate* for Salamanca (Madrid, MS 3385, f. 108r) have the same table, beginning in January, rather than in March, and there are other minor differences (Table TV 9). Since the format of this table is so unusual (i.e., the argument is the day of the year, and the entries are given in units of time), it is plausible to take it as the direct ancestor of Zacut's table for the equation of time.

AP 6. Table for the entry of the Sun into each zodiacal sign

This table has 12 columns (one for each zodiacal sign) and 136 rows (one for each year, beginning with 1473), and gives the day, the hour, and the minute of the moment when the Sun enters into each of the zodiacal signs. It is explicitly mentioned in chapter 2 of the *Ḥibbur*, and its use is explained in chapter 3 of both the Castilian and the Latin versions. In Table AP 6A we present an excerpt of the column for the entry of the Sun into the sign of Aries.

In the Castilian version two pages are missing and, as a result, there are no entries for years 103 to 136; but in chapter 3 of that version we are informed that this table was intended to cover 136 years.

Each entry differs from the previous one by 365 days and 5;49h or 5;50h (for leap years, an extra day is subtracted). The value corresponding to year 137 would then be 9d 15;40h; it differs by 1 day and 20 minutes from that for year 1. This is a parameter explicitly mentioned in chapter 2 of the *Ḥibbur*: "After these 136 years . . . , subtract for each revolution 1 day and 1/3 hour, i.e., 20 minutes" (MS B, f. 11a; MS S, f. 6v: *despues destos 136 años . . . as de menguar por cada reuolucion un dia e un 3° de hora que son 20 mjnutos*).

Table AP 6A: Entry of the Sun into the sign of Aries

Year	March
1	10d 16; 0h
2	10d 21;49h
3	11d 3;39h
4	10d 9;28h
. . .	. . .
134	9d 22;11h
135	10d 4; 0h
136	9d 9;50h

The entries in this table may have been derived from those for the daily solar positions (Table AP 2), using linear interpolation, and those in the table for the equation of time (Table AP 5), as can be seen in the following recomputation of the first value in that table, corresponding to the entry of the Sun in the sign of Aries in March 1473.

The true positions of the Sun for the dates before and after the Sun reaches Ari 0° are found in Table AP 2:

March 10, 1473 . . . Psc 29;20,55°
March 11, 1473 . . . Ari 0;20, 3°

The Sun moves 0;59,8° on that day, and thus it takes 15;52h to reach Ari 0°. The equation of time for March 10 is to be found in Table AP 5: 0;8h. Therefore, the moment sought is March 10 at 15;52 + 0;8 = 16;0h, in perfect agreement with the entry in the *Almanach Perpetuum*.

This table also allows us to derive the length of the tropical year used by Zacut (see also the comments to Table AP 4). It is easy to check that from year 1 to year 136 there is an excess of 32 days and 17;50h over a multiple of 365 days. This excess corresponds to an excess of 5;49,15,33,20h in one year, in close agreement with that given in the table of the revolutions in the Alfonsine Tables as presented in the *editio princeps* (5;49,15,43h). The value (rounded to seconds) used by Zacut for the length of the tropical year is thus 365d 5;49,16h.

AP 7. Table for the daily lunar positions

The use of this table is described in chapter 3 of the *Ḥibbur*, and in chapter 4 in both versions of the Leiria edition. The table contains three kinds of information: (i) the true longitudes of the Moon for mean noon at Salamanca for each day of a cycle of about 31 years beginning on March 1, 1473; (ii) the correction to be applied to each of the above longitudes after one cycle of about 31 years has elapsed; (iii) the corrections for the days and leap years in other cycles.

(i) The 11,325 entries for the true longitudes of the Moon are given to minutes. This period of 11,325 days is equal to 31 Julian years and 2 days, which is exactly the lunar cycle introduced by Jacob ben David Bonjorn (also called Jacob Poel), and used in his tables for true syzygies computed for Perpignan with epoch 1361.[58] In the prologue of his *Ḥibbur*, Zacut indicated that when looking for a period to compute the lunar positions in his almanac: "I did not find a more accurate period than the period found by

[58] Chabás 1991, and Chabás 1992. Samsó (1997) has recently offered evidence that Bonjorn's cycle is attested in a Maghribī source, dated approximately 1340.

R. Jacob Poel [for syzygies]" (MS B, f. 7a; MS S, f. 1v: *non hallé tienpo mas zercano e mas derecho quel tienpo que alló Ra. Jacob puel para sus conjunciones e a él do gracias porque alunbró mis ojos en este camino*; cf. Cantera 1931, p. 152). The only other medieval table with consecutive lunar positions for 11,325 days that we have found is in Vatican, MS Heb. 384, ff. 347a-359a; no headings or dates appear in that manuscript and the day count is given in decimal notation using a symbol for zero and the first nine letters of the Hebrew alphabet (to be read from left to right, rather than from right to left as is normally the case in Hebrew). We have not determined the method for computing this table or the dates for which it was computed.

As was the case with the table for the daily solar positions, the entries for the daily lunar positions in the *Almanach Perpetuum* derive from the Alfonsine Tables as presented in the *editio princeps* (Ratdolt 1483). In the following example we have again used 0;0,27,20d to account for the time difference between Salamanca and Toledo.

All quantities below are given in degrees except for the date that is given in days since the Incarnation. On the method for recomputing the entries, see Poulle 1984 (pp. 202–204).

March 1, 1473 (noon)	2,29,21,48; 0,27,20
Lunar mean motion	0,15;43, 0,15
Lunar mean center	0,54;48,40,34
Lunar mean anomaly	5, 7;54,51,35
Equation of center	+7;54,28
Provisional eq. of anomaly	+3;13,52
Diversitas diametri	+0; 0,49
Equation of anomaly	+3;30,24
True longitude of the Moon	0,19;13,24
Entry in the *Almanach Perpetuum*	0s 19;13
Difference (Zacut - Recomputation)	0; 0

We have recomputed all the entries for March 1473 and found discrepancies no greater than ± 1 minute, with the following distribution:

Z − C	$-1'$	$0'$	$+1'$
Frequency	7	18	6

It is very remarkable that the 11,325 entries in Zacut's lunar ephemeris, each one of them involving many more computations than those displayed in the above example, were calculated so carefully. It is also noteworthy that Zacut used "trigesimal" notation,[59] rather than sexagesimal notation as in the *editio princeps* of the Alfonsine Tables, to define the length of the cycle used: "12 thirds, 17 seconds, and 15 units of days" (MS B, f. 7b; MS S, f. 1v: *12. terceros 17. segundos e 15. primeros de días*), that is,

$$12 \cdot 30^2 + 17 \cdot 30 + 15 = 11{,}325\text{d}.$$

(ii) As explained in chapter 3 of the *Ḥibbur*, the true longitude of the Moon after one cycle of 31 years is obtained by adding two quantities to the longitude of the Moon listed in the table: a constant value of 182°, and a variable correction for each day of the cycle. Its values are listed here to minutes, and they range from 0;33° to 0;49°.

To recompute the entry for the correction listed in Zacut's table for March 1, one needs first to calculate the position of the Moon 11,325 days later, namely, March 3, 1504. This is accomplished by following the same procedure as in (i). Again, all quantities below are expressed in degrees except for the date that is given in days since the Incarnation.

March 3, 1504 (noon)	2,32,30,33; 0,27,20
Lunar mean motion	3,18;23,11,46
Lunar mean center	0,55;15,36,14
Lunar mean anomaly	5, 8;54,30,49
Equation of center	+7;58, 5
Provisional eq. of anomaly	+3; 9,58
Diversitas diametri	+0; 1,45
Equation of anomaly	+3;26,36
True longitude of the Moon	3,21;49,48

Thus the difference between the true longitudes for March 3, 1504 and March 1, 1473 (see above) is: 3,21;49,48° – 0,19;13,24° = 3,2;36,24°, that is, it exceeds "6 zodiacal signs and 2°" (182°) by 0;36,24° ≈ 0;36°. This is exactly the entry found in the *Almanach Perpetuum* for the correction to be applied to the true longitude of March 1, 1473 for finding that of March 3, 1504.

In order to recompute the entries for the corrections given by Zacut, we have calculated all the true longitudes of the Moon for 31 days, beginning

[59] In fact, "trigesimal" notation is a combination of base 10 and base 30. MS L (ff. 23r and 31r) also contains calendrical tables, not included in the edition of 1496, where the entries are given in "trigesimal" notation; for instance, in Table HG 63 (f. 31r) the number of days for cycle 1 in a list of 19-year lunar cycles is given as: 7,21,10 (where the digits are written in the Hebrew order, i.e., from right to left) which here means:

$$7 \cdot 30^2 + 21 \cdot 30 + 10 = 6{,}940\text{d}.$$

with March 3, 1504, and we have compared them with those for March 1473. The resulting discrepancies are no greater than ±2 minutes, and exhibit the following distribution:

Z − C	−2′	−1′	0′	+1′	+2′
Frequency	1	2	20	6	2

We therefore conclude that this was, most likely, the procedure used by Zacut to compute this correction for lunar longitudes.

(iii) The correction for the days in prior or subsequent 31-year cycles is displayed as a square matrix of 9 integers for each year. The integers in the central row are always 1, 2, and 3. Those in the first (third) row represent the number of days to be subtracted (added) to obtain the date corresponding to a year which belongs to 1, 2, or 3 cycles of 31 years before (after) the cycle beginning in 1473. The correction for the leap years is given, for each year, as an integer number between 1 and 4, after the letter "b" (for bissextile). For example, the first table (year 1473) has "b3," which means that during the third 31-year cycle after the one beginning in 1473 the year pertaining to this particular table is a leap year. Both these corrections appear in Bonjorn's tables, and Zacut has simply accepted them (Chabás 1992, pp. 73–75).

AP 8. Table for syzygies

In the Castilian version (as well as in chapter 5 of the Latin version), we are told that in constructing his table of syzygies Zacut followed "el Poel," i.e., Jacob ben David Bonjorn, also known as Jacob Poel. Indeed, this table preserves the essential features of Bonjorn's table for syzygies.[60] It contains two kinds of information: (i) the moment of each true syzygy in a cycle of 31 years and 2 days, beginning in March 1478 (in Bonjorn's tables, the 31-year cycle begins in March 1361); (ii) the correction to be applied to the time of each syzygy given previously in order to find the time of true syzygy in cycles other than that covered by the period 1478–1508.

(i) More precisely, the length of Bonjorn's 31-year cycle is 767 syzygies, which is a non-integer number of synodic months (383.5); this is the number of true syzygies listed by Zacut in his table. For each of them we are given the date (year, month, day, and weekday), and the time (hour and minute) at which it takes place.

It should be pointed out that Zacut's list of 767 true syzygies begins in March 1478, whereas his tables for the daily lunar and solar positions begin in March 1473.

[60] Albuquerque 1986, p. 82; see also Chabás 1992, pp. 226–241.

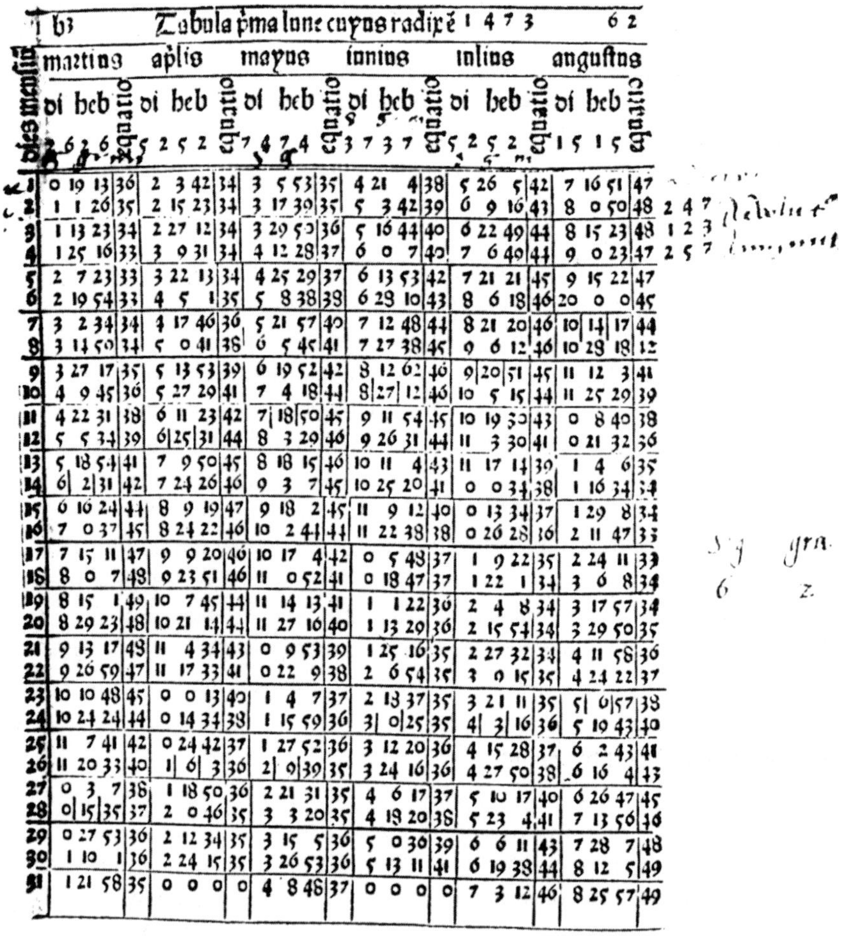

b3 Tabula prima lune cuyus radix e 1 4 7 3 62

dies mensis	martius	equatio	aprilis	equatio	mayus	equatio	iunius	equatio	iulius	equatio	augustus	equatio
	di heb		di heb		di heb		di heb		di heb		di heb	
	2 6 2 6		5 2 5 2		7 4 7 4		3 7 3 7		5 2 5 2		1 5 1 5	
1	0 19 13	36	2 3 42	34	3 5 53	35	4 21 4	38	5 26 5	42	7 16 51	47
2	1 1 26	35	2 15 23	34	3 17 39	35	5 3 42	39	6 9 16	43	8 0 50	48
3	1 13 23	34	2 27 12	34	3 29 50	36	5 16 44	40	6 22 49	44	8 15 23	48
4	1 25 16	33	3 9 31	34	4 12 28	37	6 0 7	40	7 6 40	44	9 0 23	47
5	2 7 23	33	3 22 13	34	4 25 29	37	6 13 53	42	7 21 21	45	9 15 22	47
6	2 19 54	33	4 5 1	35	5 8 38	38	6 28 10	43	8 6 18	46	20 0 0	45
7	3 2 34	34	4 17 46	36	5 21 57	40	7 12 48	44	8 21 20	46	10 14 17	44
8	3 14 50	34	5 0 41	38	6 5 45	41	7 27 38	45	9 6 12	46	10 28 18	12
9	3 27 17	35	5 13 53	39	6 19 52	42	8 12 62	46	9 20 51	45	11 12 3	41
10	4 9 45	36	5 27 29	41	7 4 18	44	8 27 12	46	10 5 15	44	11 25 29	39
11	4 22 31	38	6 11 23	42	7 18 50	45	9 11 54	45	10 19 30	43	0 8 40	38
12	5 5 34	39	6 25 31	44	8 3 29	46	9 26 31	44	11 3 30	41	0 21 32	36
13	5 18 54	41	7 9 50	45	8 18 15	46	10 11 4	43	11 17 14	39	1 4 6	35
14	6 2 31	42	7 24 26	46	9 3 7	45	10 25 20	41	0 0 34	38	1 16 34	34
15	6 16 24	44	8 9 19	47	9 18 2	45	11 9 12	40	0 13 34	37	1 29 8	34
16	7 0 37	45	8 24 22	46	10 2 41	44	11 22 38	38	0 26 28	36	2 11 47	33
17	7 15 11	47	9 9 20	46	10 17 4	42	0 5 48	37	1 9 22	35	2 24 11	33
18	8 0 7	48	9 23 51	46	11 0 52	41	0 18 47	37	1 22 1	34	3 6 8	34
19	8 15 1	49	10 7 45	44	11 14 13	41	1 1 22	36	2 4 8	34	3 17 57	34
20	8 29 23	48	10 21 14	44	11 27 16	40	1 13 29	36	2 15 54	34	3 29 50	35
21	9 13 17	48	11 4 34	43	0 9 53	39	1 25 16	35	2 27 32	34	4 11 58	36
22	9 26 59	47	11 17 33	41	0 22 9	38	2 6 54	35	3 0 15	35	4 24 22	37
23	10 10 48	45	0 0 13	40	1 4 7	37	2 13 37	35	3 21 11	35	5 0 57	38
24	10 24 24	44	0 14 34	38	1 15 59	36	3 0 25	35	4 3 16	36	5 19 43	40
25	11 7 41	42	0 24 42	37	1 27 52	36	3 12 20	36	4 15 28	37	6 2 43	41
26	11 20 33	40	1 6 3	36	2 9 39	35	3 24 16	36	4 27 50	38	6 16 4	43
27	0 3 7	38	1 18 50	36	2 21 31	35	4 6 17	37	5 10 17	40	6 26 47	45
28	0 15 35	37	2 0 46	35	3 3 20	35	4 18 20	38	5 23 4	41	7 13 56	46
29	0 27 53	36	2 12 34	35	3 15 5	36	5 0 30	39	6 6 11	43	7 28 7	48
30	1 10 1	36	2 24 15	35	3 26 53	36	5 13 11	41	6 19 38	44	8 12 5	49
31	1 21 58	35	0 0 0	0	4 8 48	37	0 0 0	0	7 3 12	46	8 25 57	49

Figure 8. Table AP 7: Table for the daily lunar positions, first page only, *Almanach Perpetuum*, Leiria, 1496, f. 30r (Madrid, Biblioteca Nacional, shelfmark I-1077).

We have recomputed all the syzygies in Zacut's list using Bonjorn's tables for Perpignan, following the method described in Bonjorn's canons, as Zacut could have done. For example, the first syzygy in Zacut's table occurs at 6;0h (Salamanca time) on March 4, 1478. This syzygy is distant in time from a syzygy listed in Bonjorn's tables by exactly 3 Bonjorn's cycles, and occurred at 7;39h (Perpignan time) on February 25, 1384. The correction for this particular syzygy is also found in Bonjorn's tables: 0;24h. Bonjorn's method for computing a future syzygy is simply to subtract the correction from the time of the given syzygy (7;39h) as many times as cycles have elapsed ($3 \cdot 0;24$h in this case). The result is 6;27h. Now, if we allow a difference in longitude between Perpignan (32;30°) and Salamanca (25;46°)[61] of 6;44°, that is, a time difference of 0;27h, the result is 6;0h, in exact agreement with the entry given by Zacut. The distribution of the results (in minutes of time) follows:

Z − C	≤ -6	$[-5, -3]$	-2	-1	0	$+1$	$+2$	$[+3, +5]$	$\geq +6$
Frequency	9	22	12	39	614	39	7	13	12

This is a satisfactory result globally, for it shows that Zacut correctly calculated more than 90% of the entries with an accuracy of ± 1 min. Some of the discrepancies may be due to copyist errors in the manuscript(s) of Bonjorn's table used by Zacut.

(ii) The 767 values for the corrections listed by Zacut allow us to calculate the time of the true syzygies that take place one 31-year cycle later or earlier, by subtracting or adding the corresponding correction. When seeking the time of a true syzygy for still more distant cycles, the same amount has to be subtracted as many times as the number of cycles that have elapsed. All these corrections are directly copied from Bonjorn's tables, but for one feature. Bonjorn's corrections are given in minutes of time and sub-multiples of them, called "parts," such that 17 parts = 1 min; it is a rather peculiar sub-unit, but it has an internal explanation (Chabás 1992, pp. 77–80). However, the corrections listed by Zacut are only given to minutes, and they are obtained by rounding those in Bonjorn's tables. The agreement of Zacut's corrections with those of Bonjorn, as rounded, is exact in all but 39 out of 767 cases.

AP 9. Table for the mean motion in lunar anomaly

This table displays the mean motion in lunar anomaly, to minutes of arc for each year in a period of 180 years, for each month of the year, and

[61] See Table AP 49.

for each day, hour, and minute. As far as we know, 180 years as a period for lunar anomaly is only used in two other tables of this kind, namely, the Almanac of Azarquiel (Millás 1943–1950, p. 170), and the *Tabule Verificate* for Salamanca (see Table TV 5, above). Actually, Table AP 9 shares all features, and parameters, with Table TV 5.

In our table, the entry for year 1 is 9s 24;51°. According to the table for the mean motion of lunar anomaly in the *editio princeps* of the Alfonsine Tables, its value at noon on the day before March 1, 1473 at Toledo is 294;45,0°. Since the lunar anomaly has progressed 0;5,57° from noon at Toledo to noon at Salamanca, we obtain exactly the entry given by Zacut in his table. This is, therefore, another value of Alfonsine origin. Moreover, the line-by-line differences in the table are 88;43° or 88;44° (for non-leap years), and 101;47° or 101;48° (for leap years), plus 13 revolutions in each case. These values are in agreement with the Alfonsine parameter for the mean motion in lunar anomaly: 13;3,53,57,30,21,4,13°/d.

AP 10. Table for the equation of center for the Moon

The heading of this table is *equatio argumenti*. The entries are given in minutes and seconds as a function of the number of days after mean syzygy.

As the table shows, the tabulated function, despite its heading, is the equation of center, i.e., the correction to be applied to the mean lunar anomaly to obtain the true lunar anomaly. In medieval Latin astronomical literature this correction is usually called *equatio centri*. The values given here agree with those in the table for the *equatio centri* given as a function of the double elongation in the *editio princeps* of the Alfonsine Tables. For example, the value for the elongation corresponding to 1 day in the *editio princeps*, as rounded, is 12;11,27°. The double elongation is therefore 24;22,54°. By interpolation between the entries for 24° (+3;31°) and 25° (+3:40°) in the *editio princeps*, we obtain +3;34°, in agreement with the entry in Zacut's table.

Table AP 10A: The Equation of Center for the Moon

Days	*Eq. arg.*	Days	*Eq. arg.*
1	+ 3;34°	9	−11; 6°
2	+ 7; 5°	10	−13; 8°
3	+10;14°	11	−12; 9°
4	+12;34°	12	− 9;33°
5	+13; 1°	13	− 6;16°
6	+10; 8°	14	− 2;45°
7	+ 3;18°	15	+ 0;51°
8	− 5; 9°		

AP 11. Table for the motion of the lunar nodes

The motion of the lunar nodes is presented in two tables that are similar to one another, the headings of which are *Tabula veri motus capitis draconis in 903 annis*, and *Tabula medii motus capitis draconis in 903 annis*. The second table gives the longitude of the ascending node, whereas the first one displays its complement in 360°. In both cases, there are entries given to minutes for each year in a period of 93 years, for each month of the year beginning in March, and for each day up to 30 days. The motion of the ascending node is retrograde, but many medieval tables display an increasing longitude for the ascending node with the instruction to subtract the tabulated value from 360°. Examples of this usage are found in the *Handy Tables* (Stahlman 1959, pp. 107, 243–248), al-Battānī's zij (Nallino 1899–1907, 2:19–23, 204), the Toledan Tables (Toomer 1968, p. 50), and Levi ben Gerson's tables (Goldstein 1974, pp. 172–181). This convention for the lunar node is also followed in the *editio princeps* of the Alfonsine Tables (see Poulle 1984, p. 139); the instruction for subtracting the tabulated value from 360° in John of Saxony's canons is found in chapter 19 (Poulle 1984, p. 73).

Note that the headings in both the Castilian and the Latin versions of the *Almanach Perpetuum* share the same error: 903 years instead of 93 years, the latter being the period for the nodes in previous almanacs.[62] This is not the only error shared by both Leiria versions, for in the column for the years in the second table we find "26 years" instead of "62 years."

The entry for the longitude of the ascending node for year 1 is 4s 5;49°. As was the case for the radices previously examined, this radix can also be derived from the Alfonsine Tables: according to the table for the mean motion of the lunar nodes in the *editio princeps*, the value for noon on the day preceding March 1, 1473 at Toledo is 125;48,34°. The difference in longitude between Toledo and Salamanca adds just 1 second. Hence, with rounding we obtain exactly the entry in Zacut's table. This table is the complement in 360° of the corresponding table in the *Tabule Verificate* (Madrid, MS 3385, f. 109r: Table TV 11), beginning in March 1461 rather than in March 1473: the entry for year 13 in the Madrid manuscript is 7s 24;11°, the complement in 360° of the entry for year 1 in the *Almanach Perpetuum*. The remaining entries are straightforward.

The mean motion of the lunar nodes embedded in this table also derives from the corresponding Alfonsine parameter: –0;3,10,38,7,14,49,10°/d.

Thus, all the material in Zacut's tables for the lunar mean motions and equations derive from the Alfonsine Tables, although presented somewhat differently.

[62] See, e.g., the Almanac of Azarquiel in Millás 1943–1950, p. 175. Note that Table HG 12, corresponding to Table AP 11, has "93 years" in its headings.

AP 12. Table for the solar elongation from the lunar node

The heading for this table is *Tabula motus solis a capite draconis in 56 annis*. It lists the differences in longitude between the Sun and the lunar ascending node for each year in a period of 56 years, and for each hour in a day. The entries are given in signs, degrees, and minutes. The purpose of this table for the elongation of the Sun from the lunar node is to call attention to eclipse possibilities: when the distance is such that, at a syzygy, the Sun lies in the nodal zone, it is appropriate to compute the circumstances of a lunar or solar eclipse. It is unusual to find a table with this argument, and we know of only two such tables before Zacut: Table TV 10 that is very close to Table AP 12; and one by Ben Verga that has the heading, "Table for finding the motion of the Sun from the ascending node." The entries in Ben Verga's table are given for epoch 1400, and then at intervals of 28 years, single years, months (beginning in January), and days (Paris, MS Heb. 1085, f. 95b; Oxford, MS Poc. 368 [Nb. 2044], f. 231b). The underlying parameters are slightly different for, in Ben Verga's table, the entry for 28 years is 6s 1;32°, whereas in Table AP 12 the corresponding value for 28 years is 6s 1;46° (computed from the entries for year 29 and year 1).

The entry in Table AP 12 for year 1 is 3s 25;16°. This value can be deduced from previous tables in the *Almanach Perpetuum*. If we take the longitude of the true Sun for noon of March 1, 1473 from Table AP 1 (Psc 20;26,30°), and subtract the arc traversed by the Sun in one day (0;59,8°), we obtain 11s 19;27,22° for the solar longitude at noon of the previous day. We then subtract the position of the lunar node given in Table AP 11 for year 1 (7s 24;11°), and we obtain exact agreement with the entry in Zacut's table. This table could have also been partially copied from the corresponding Table TV 10 in the *Tabule Verificate*, beginning in March 1461 rather than in March 1473: the entry for year 13 in the Table TV 10 is 3s 25;16°, the entry for year 1 in the *Almanach Perpetuum*. The remaining entries are straightforward.

A separate sub-table displays entries, also given in signs, degrees, and minutes, for each day in a year beginning in March. The entry for March 1 is 0s 1;3°. This is the same sub-table as that in the *Tabule Verificate* (Table TV 10).

AP 13. Table for the equation of eclipses and parallax of the Moon

The heading for the table, presented in 3 columns and reproduced as Table AP 13A, is *Ta. eqt. eclip. & div. aspec. lune*.

The purpose of col. 3, where the entries are given in minutes of arc, is to correct the parallax (see Table AP 14) when the Moon is not at syzygy. The argument, col. 1, is the double elongation, given here for each integer

Tabula motus solis acapite draconis in 56 annis								Motus solis acapite in horis			Ta.eq̃t.eclip.z dix.aspec.lune		
anni	s̃	g̃	m̃	anni	s̃	g̃	m̃	hore	g̃	m̃	signa argumēt	minuta argum	m̃ dix aspec
1	3	25	16	29	9	27	2	1	0	3			
2	4	14	20	30	10	16	7	2	0	5			
3	5	3	26	31	11	5	13	3	0	8			
4	5	23	34	32	11	25	21	4	0	10			
5	6	12	40	33	0	14	25	5	0	13			
6	7	1	44	34	1	3	31	6	0	15			
7	7	20	50	35	1	22	37	7	0	18	1	5	3
8	8	10	58	36	2	12	45	8	0	20	2	3	9
9	9	0	3	37	3	1	49	9	0	23	3	0	17
10	9	19	8	38	3	20	55	10	0	25	4	2	26
11	10	8	14	39	4	10	1	11	0	28	5	4	30
12	10	28	22	40	5	0	9	12	0	31	6	5	32
13	11	17	26	41	5	19	13	13	0	33	7	4	30
14	0	6	32	42	6	8	19	14	0	36	8	2	26
15	0	25	38	43	6	27	25	15	0	38	9	0	17
16	1	15	46	44	7	17	33	16	0	41	10	3	9
17	2	4	50	45	8	6	37	17	0	43	11	5	3
18	2	23	56	46	8	25	43	18	0	46	12	5	0
19	3	13	2	47	9	14	49	19	0	48			
20	4	3	10	48	10	4	57	20	0	51			
21	4	22	14	49	10	24	1	21	0	54			
22	5	11	20	50	11	13	7	22	0	57			
23	6	0	26	51	0	2	13	23	0	59			
24	6	20	34	52	0	22	21	24	1	2			
25	7	9	38	53	1	11	25						
26	7	28	44	54	2	0	31						
27	8	17	50	55	2	19	37						
28	9	7	58	56	3	9	45						

Figure 9. Table AP 12: Solar elongation from the lunar node, *Almanach Perpetuum*, Leiria, 1496, f. 63r (Madrid, Biblioteca Nacional, shelfmark I-1077).

Table AP 13A: Equation of eclipses and parallax

[1] *signa argument*	[2] *minuta argum*	[3] *m div aspec*
1	5	3
2	3	9
3	0	17
4	2	26
5	4	30
6	5	32
7	4	30
8	2	26
9	0	17
10	3	9
11	5	3
12	5	0

zodiacal sign. Col. 3 is a short version of a column that already appears in Ptolemy's *Handy Tables* (Stahlman 1959, p. 257, col. 4), where the argument is displayed at intervals of 6°, within a table called "table of corrections." In fact, Ptolemy's table is also found in many sets of tables: al-Battānī's zij (where this column is labeled "eccentric"), the Almanac of Azarquiel (where it is headed el *centro salient*, and given as "seconds" instead of "minutes"), the Toledan Tables (where we find *circulus egressus* as the heading for this column) and the *editio princeps* of the Alfonsine Tables (where this column is also headed *circulus egressus*, and the table is called *tabula attacium*).[63]

One would expect col. 2 to be based on Ptolemy's "table of corrections" as well, that is, the column whose purpose is to correct the parallax when the Moon is not at apogee, and headed *circulus brevis* or "epicycle" in subsequent literature (Stahlman 1959, p. 257, col. 3): see Table AP 13B, where the argument is the lunar anomaly. For an explanation of the way the entries in Ptolemy's table can be computed, and their use for correcting components in parallax, see Rome (1931, pp. 52–55).

Now, chapter 6 of the *Ḥibbur* gives a very confused description of this table (MS B, f. 21a):

[63] For al-Battānī, see Nallino 1899–1907, 2:89; for the Almanac of Azarquiel, see Millás 1943–1950, p. 233; for the Toledan Tables, see Toomer 1968, pp. 116–117; for the Alfonsine Tables, see the 1483 edition (f. 18v). Note that the Arabic *al-taqwīm* (transliterated here as *attacium*, probably from *attacuim*) can mean "correction" or "equation," as noted by Nallino 1899–1907, 2:350.

Table AP 13B: Ptolemy's table for correcting parallax

[1] Lunar anomaly	[2'] Correction (min)
1	1
2	3
3	6
4	9
5	11
6	12
7	11
8	9
9	6
10	3
11	1
12	0

Enter the small table made for correcting the eclipses with the signs [MS L, f. 207r, adds: of the anomaly], and take the minutes opposite them. They are in proportion to 60, [MS B mg, and MS L add: and take from the digits, and similarly from the minutes of an hour, the proportion to 60,] and add the resulting proportion to the digits and similarly for the minutes [of time] that you took from the table of solar eclipses if the anomaly is from 3s to 9s. But subtract the resulting proportion if the anomaly is from 9s to 3s (MS S, f. 22r: *entra en una tabla pequeña ques fecha para precindir o estatimar los eclipses con los signos del argumento . . . e lo que uiniere de aquella proporcion añadelo: sobre los digitos e asi mesmo haras de los mjnutos de hora que tomaste en la tabla del eclipse del sol. si el argumento era de .3. signos hasta .9. o menguaras esta dicha proporcion si el argumento fuere desde 9. signos asta .3. signos*; cf. Cantera 1931, pp. 178–179).

What can be inferred from this passage is that all entries in col. 2 are positive for values of the lunar anomaly from 3s to 9s, and the rest negative. When subtracting the entries derived from Ptolemy's table in col. 2' from the entries in col. 2, with their proper algebraic signs, we obtain almost constant differences. This leads us to believe that column 2 in the table of the *Almanach Perpetuum* was intended to represent the first correction to parallax (ultimately in Ptolemy's table), but that some miscopying or shifting of the entries in this table has made it useless.

In the *Ḥibbur* (MS L, f. 139r), this table is called "table for the correction of eclipses and parallax" (Table HG 17). Col. 1 has the heading: "signs of the center and of the anomaly"; col. 2 has the heading: "minutes of the correction for anomaly"; and col. 3 has the heading: "minutes of parallax for the center." The term "center" here refers to the double elongation. The entries in all columns of this table in this Hebrew manuscript agree exactly with those in the Leiria edition.

Column 2 seems to derive from Table TV 17 in Madrid, MS 3385.

AP 14. Table for parallax

This is the same table as TV 12, but for some features: for instance, the sub-table for Capricorn is missing in both Leiria versions, and the reported time of nonagesimal for Leo is −0;24h instead of 0;24h.

The parallax table in the *Almanach Perpetuum* contains quite a number of errors (indeed, a whole sub-table is missing!), and it is thus very difficult to derive the underlying parameters. The best fit for the entries of parallax in latitude occurs for an obliquity, ε, of 23;33° and a horizontal parallax, p_0, of 0;53,20°, which is the value used by Levi ben Gerson (Goldstein 1974, p. 116). Surprisingly, in the case of the parallax in longitude, the best results are obtained with $\varepsilon = 23;33°$ and $p_0 = 0;57,30°$ (Chabás, Roca, and Rodríguez 1988, pp. 237–248). The use of these two different values of p_0 in two different columns of the same table seems to point towards Jacob ben David Bonjorn, whose tables for parallax share the same peculiarity (Chabás 1992, pp. 111–112). We suspect that Zacut modeled his tables of parallax for the latitude of Salamanca on those calculated by Bonjorn (called "Poel" by Zacut: see the text below) for Perpignan, that in turn depend on those of Levi, but we cannot explain how this was done.

The canons of the *Almanach Perpetuum* are silent on the way this table was computed, and on the parameters on which it is based. This is also the case in the *Ḥibbur*, but in chapter 4 we find a discussion of that issue when Zacut explains his procedures for correcting lunar latitude due to parallax (MS B, f. 15b:5 ff; see also MS S, f. 14r-v):[64]

> Know that according to the opinion of some recent [scholars], the latitude of the Moon is not greater than 4;29°, and they call this latitude the exact latitude. On the basis of this latitude Poel, may his memory be for a blessing, established [his table for] eclipses, and I have followed him in this respect. Here R. Judah ben Asher[65] was in error in the second procedure that he gave for finding the parallax (*ḥilluf ha-habbaṭa*) in the second table for the parallax (*mar'eh*)[66] in latitude, i.e., table 71 among

---

[64] This passage is discussed in Goldstein and Chabás 1999.

[65] Judah ben Asher (d. 1391) is not to be confused with his namesake who died in 1349: see Alfred Freimann 1920, pp. 149–156.

[66] The usual expressions in Hebrew for parallax are *ḥilluf ha-habbaṭa* (Goldstein 1974, p. 268) and *shinui ha-mar'eh* (see, e.g., Judah ben Asher in Vatican, MS Heb. 384, 322b:23). In some instances Zacut has used only one of the two words in each expression.

his tables is established on the basis of exact latitude, and it is identical with Ptolemy's table for each one [of them] according to his opinion for the latitude, for the parallax (*ḥilluf*) corresponding to 90° is 0;48,45° and it is equal [or, equivalent] to 0;53,34°[67] of Ptolemy. And he established the eclipses for each one of the procedures using the [MS L adds: table of] latitude in Ptolemy which is 5°. Anyone who examines the tables that recent [scholars] have made will understand this. Moreover, he [Judah ben Asher ?] said that it concerns the mean [MS L adds *supra*: that is, mean distance], but this is not so, for it is only established according to the apogee of the epicycle, that is, table 71 among his tables. Therefore, if you wish to know the exact latitude of the Moon from this table after finding the latitude, subtract a tenth of the latitude and the remainder is the exact latitude of the Moon.[68] Also, others said that it [the maximum lunar latitude] is 4;45°, and for them [i.e., those holding this opinion] a twentieth is subtracted. R. Judah ben Asher also mentioned another opinion, [namely] that the anomaly also introduces [a component of] latitude, as is the case for the five planets. According to that opinion, this [component] of latitude is to the south from 3s to 9s, and to the north in the upper half [of the epicycle]. This component of latitude at the epicyclic apogee and perigee[69] is 0;12, and for 1s, 5s, 7s and 11s its latitude is 0;10, and for 2s, 4s, 8s, 10s its latitude is 0;6, and for 3s and 9s, i.e., mean

---

[67] MS B: 0;53,34°, corrected from 0;53,33°; MS L: 0;53,34°; MS S: 0;51,34°.

[68] The maximum lunar latitude is 5°, and the maximum exact lunar latitude is 4;30°; hence the difference between them is 0;30°. This can be expressed as

$$1/10 \cdot 5^\circ = 1/10 \cdot 300' = 30'.$$

In the text we are told, in general, that

$$\beta - 1/10 \cdot \beta = \beta',$$

where β' is the exact lunar latitude because

$$\beta'/\beta = 4;30/5;0$$

or

$$(\beta - \beta')/\beta = 0;30/5;0.$$

[69] Note that this second component of latitude, due to the anomaly, is based on the following formula:

$$\beta_2 = 0;\ 12 \cdot \cos(\alpha).$$

Thus,

$$0;12 \cdot \cos 30^\circ = 0;10,$$

$$0;12 \cdot \cos 60^\circ = 0;\ 6,$$

and

$$0;12 \cdot \cos 90^\circ = 0;\ 0.$$

distances, there is no latitudinal [component] as is the case for Saturn, Jupiter, and Mars.

Note that there is neither a "table 71," nor a parallax table in the extant defective copy of the astronomical work by Judah ben Asher that is uniquely preserved in Vatican, MS Heb. 384; nevertheless, there is a passage in this manuscript that has 0;48,45° and seems to be the value for the horizontal lunar parallax (f. 327a). Cantera, or perhaps Juan de Salaya, the translator into Castilian of the *Ḥibbur*, mistakenly took this value to be 0;43,45°. The value attributed to Ptolemy is correct, and it is the lunar parallax at the greatest distance for an argument of 90° as found in the *Almagest* (V, 18, col. 3).

This text seems to suggest that Zacut interprets the value 0;48,45° to apply to the apogee of the epicycle at syzygy, whereas Judah ben Asher took it to refer to mean distance at syzygy. Could this mean that Zacut first accepted the view he ascribes to Judah ben Asher (i.e., in 1476, at the time of the occultation of Venus by the Moon, when writing the gloss at the end of chapter 14), and later rejected it (when writing chapter 4)?

The parameter, 4;29°, for the maximum lunar latitude is also found in Table AP 18, and it is well attested (see, e.g., Chabás and Goldstein 1994, p. 22), but not 4;45°. The formula, $\beta_2 = 0;12 \cdot \cos(\alpha)$, for the second component of lunar latitude, is found in the text of Judah ben Asher (Vatican, MS. Heb. 384, f. 300b):

> And some recent [scholars] have argued that there is another latitude for the Moon because its epicycle is inclined to the great orb [i.e., the deferent] to the north and to the south. The way to correct it is as follows: take the anomaly of the Moon at the given time and then multiply its cosine by 12 and divide it by 60. The result is the latitude of the Moon due to its anomaly. Note: if the anomaly lies between 0[s] 0[°] and 1[s] 30[°] or between 4[s] 30[°] and 6[s] 0[°], then this [component of] latitude is to the north. But if the anomaly lies between 1[s] 30[°] and 4[s] 30[°], this [component of] latitude is to the south. Then note: if the first latitude and the second both lie in the same direction, add them together, and it is the latitude of the Moon; but if one is to the north and the other to the south, subtract the smaller from the larger, and the remainder is the latitude of the Moon, and its direction is the direction of the greater latitude. [end chap. 16]

Note that Judah ben Asher uses signs of 60°, whereas Zacut uses signs of 30°.

In the canons to Judah ben Asher's tables, chapter 40, there is a long passage concerning lunar parallax in longitude and latitude in which tables 71 and 72 are described (ff. 326a-327a). This is almost certainly the passage to which Zacut alludes.

AP 15. Table for solar eclipses

The subject of solar eclipses is treated extensively in chapter 8 of the Castilian version of the *Almanach Perpetuum*, and in the first part of chapter 6 in the Latin version. The same subject is covered in chapter 6 of the *Ḥibbur*.

In fact, there are 2 tables for solar eclipses: (i) a table that gives the magnitude and the half-duration of the eclipse at mean distance, and (ii) another that gives the eclipsed fraction of the solar disk.

(i) The argument for the first table is the difference between the latitude and the parallax in latitude of the Moon, $\beta - p_\beta$, given in minutes of arc for every integer from 0 to 28. Just after the "28" another "28" is (erroneously) listed, instead of 0;28,23°, which is the value mentioned in the text as the maximum argument for the visibility of a solar eclipse, as well as in the solar eclipse table in Lyon, MS Heb. 14, f. 138v (Table HG 19). As a matter of fact, in the corresponding chapter of the Castilian version we find *28 minuta e grados* (sic) *23*: the Latin version simply reads "28 23," and the text of the *Ḥibbur* has the correct value in the correct units. The magnitude of the eclipse is given in digits and minutes of a digit, and the half-duration of the eclipse in minutes of time.

To recompute this table there is an important clue in a passage in chapter 6 of the *Ḥibbur*, not included in the *Almanach Perpetuum*: "and in all this you will see that I depended on the eclipses that R. Jacob Poel computed

Table AP 15A: Solar Eclipses

Argument	Magnitude	Half-duration	Argument	Magnitude	Half-duration
0	11;47	55	15	5;33	47
1	11;22	55	16	5; 8	45
2	10;57	55	17	4;43	44
3	10;32	55	18	4;18	42
4	10; 7	55	19	3;53	40
5	9;42	54	20	3;28	38
6	9;17	54	21	3; 4	37
7	8;53	54	22	2;29**	35
8	8;28	53	23	2;14	32
9	8; 3	53	24	1;49	29
10	7;38	52	25	1;24	26
11	7;13	51	26	0;59	21
12	6;48	50	27	0;34	13
13	6;23	49	28	0; 9	3
14	5;58	48	28*	0; 0	0

* *Sic*, instead of 28;23. ** *Sic*, instead of 2;39.

except that they are not ordered according to his method, for this table is useful for you universally" (MS B, f. 19a; MS L, f. 208r; MS S, f. 19v: *e en todo esto hallaras que me arimé a la orden de los eclipses que hizo Ra. Jacob puel, aunque no uan ordenados segun el los ordeno*; cf. Cantera 1931, p. 175). But among the tables compiled by Jacob ben David Bonjorn there is no such table for solar eclipses, although there is one with a rather different pattern: it is a double argument table where the entries are given as a function of the argument of lunar latitude, ω, from 0° to 17° at intervals of half a degree, and the parallax component in latitude, p_β, from 0;6° to 0;51° at intervals of 0;3°. Each entry gives the magnitude of the eclipse, D, and its half-duration, t (Chabás 1992, pp. 249–250).

It is possible, however, to reconstruct Zacut's table from Bonjorn's according to a method that we will illustrate with the following example. Consider the case when the argument is $\beta - p_\beta = 27$ min. In Bonjorn's double argument table for solar eclipses, when $\omega = 0°$ (hence $\beta = 0°$) and $p_\beta = 0;27°$, that is, when $|\beta - p_\beta| = 0;27°$, we find D = 0;34d and t = 0;13h. These are exactly the same entries as those in Zacut's table. There are seven other pairs of values for D and t that can be taken directly from Bonjorn's table: those corresponding to $\omega = 0°$ and, successively, $p_\beta =$ 0;6°, 0;9°, 0;12°, 0;15°, 0;18°, 0;21° and 0;24°. In those cases, $|\beta - p_\beta| =$ 0;6°, . . . , 0;24°, and the corresponding values in Zacut's table are identical with those in Bonjorn's.

For the remaining entries in the column for the magnitude of the eclipse, we believe Zacut proceeded by linear interpolation between the values taken directly from Bonjorn's table. This is a legitimate procedure since D is a linear function of $|\beta - p_\beta|$:

$$D = \left(r + s - |\beta - p_\beta|\right) \cdot 6/s, \qquad [2]$$

where r and s are the radii of the Moon and the Sun, respectively, at mean distance (Chabás 1992, p. 127). The procedure of interpolation gives results in full agreement with Zacut's entries and, in particular, it implies D = 2;39d (not 2;29d as displayed in the table) when $|\beta - p_\beta| = 0;22°$. Note also that the extreme value of the argument, 0;28,23°, also derives from Bonjorn, and ultimately from Levi ben Gerson, upon whose tables for solar eclipses (Goldstein 1974, pp. 128–131) Bonjorn depended. For Levi, the lunar radius is constant at syzygies (r = 0;13,55,30°) and the solar radius at mean distance is s = 0;14,27,45°. Under those circumstances, when the eclipse is no longer visible, and D = 0d, then

$$|\beta - p_\beta| = r + s = 0;\,28,\,23,\,15°,$$

in full agreement with the entry in Zacut's table.

As for the recomputation of the remaining entries in the column for the half-duration of the solar eclipse, linear interpolation is no longer possible, for t is a function of the argument given by the equation:

$$t = [(r + s)^2 - (\beta - p_\beta)^2]^{1/2}/v, \qquad [3]$$

where t, the half-duration, is given in hours, and v, the mean motion in elongation, in degrees per hour (Chabás 1992, p. 131). We believe that, in this case, Zacut had to choose his values from entries in Bonjorn's table other than those for which $\omega = 0°$, as illustrated in the following example. Consider the case when the argument is $\beta - p_\beta = 26$ min. In Bonjorn's double argument table, when $\omega = 1;30°$ (corresponding to $\beta = 0;7°$, using the table for lunar latitude in the vicinity of $\omega = 0°$: see Table AP 18, below) and $p_\beta = 0;33°$, that is, when $|\beta - p_\beta| = 0;26°$, we find t = 0;21h, and this is exactly the entry in Zacut's table.

(ii) The second table for solar eclipses gives the area digits (with a maximum of 12) of the eclipsed solar disk as a function of the linear digits (with a maximum of 12) of the eclipsed solar diameter. It is the same as Table TV 13. This table is already found in Ptolemy's *Almagest* (VI, 8), in the *Handy Tables* (Stahlman 1959, p. 258), as well as in a number of earlier sets of tables used in medieval Spain.

AP 16. Table for lunar eclipses

Lunar eclipses are treated extensively in an unnumbered chapter just after that devoted to solar eclipses in the Castilian version, and in the second part of chapter 6 in the Latin version. The same subject is covered in chapter 7 of the *Ḥibbur*.

As was the case for solar eclipses, there are 2 tables for lunar eclipses: (i) a table that gives the magnitude, the half-duration of the eclipse, and the half-duration of totality at mean distance, and (ii) another that gives the eclipsed fraction of the lunar disk.

(i) The argument for the first table is the argument of lunar latitude, given in minutes of arc, from 5s 18;0° or 11s 18;0° to 6s 12;0° or 0s 12;0°, at intervals of 0;30°. For each value of the argument of lunar latitude we are given the magnitude of the eclipse (digits and minutes of a digit), the half-duration of the eclipse (hours and minutes), and the half-duration of totality (minutes of time). It should be noted that all the entries in the Leiria edition are shifted upwards two lines.

Zacut copied his table from Jacob ben David Bonjorn's table for lunar eclipses, as Zacut himself acknowledges in chapter 7 of the *Ḥibbur*: "And this table [for lunar eclipses] is according to the opinion of R. Jacob Poel" (MS B, f. 21b; MS L, f. 206v; MS S, f. 23v: *E esta tabla [del eclipse de la luna] es a la opinion del puel*; cf. Cantera 1931, p. 180). The table was reproduced by Cantera (1935, p. 136) and its mathematical content explained by Chabás (1992, pp. 134–144); a summary of this explanation follows.

The magnitude, D, can be recomputed by means of the following equation:

$$D = (z + r - \beta / \cos i) \cdot 6/r, \quad [4]$$

where z and r are the radii of the Earth's shadow and the Moon, respectively, at mean distance, i is the inclination of the lunar orb, and β is the lunar latitude corresponding to the argument of lunar latitude, ω, according to the expression:

$$\beta = \arctan(\tan i \cdot \sin \omega) \quad [5]$$

or by means of a table, such as Table AP 18 (see below). By taking i = 4;30° and Levi ben Gerson's values for the radii (r = 0;13,56°, z + r = 0;56,12,30°: see Goldstein 1974, pp. 123–128), we obtain results in good agreement with the entries in Bonjorn's (or Zacut's) table.

The half-duration of the eclipse, t, can be recomputed as follows:

$$t = \left[(z + r)^2 \cdot \cos^2 i - \beta^2\right]^{1/2} / v, \quad [6]$$

where v is the mean motion in elongation.

The half-duration of the totality, t′, can be recomputed as follows:

$$t' = \left[(z - r)^2 \cdot \cos^2 i - \beta^2\right]^{1/2} / v. \quad [7]$$

When we apply Levi's value for mean velocity (v = 0;30,37°) in the last two expressions, we find good agreement with Bonjorn (Zacut).

(ii) The second table for lunar eclipses gives the area digits (with a maximum entry of 12) of the eclipsed lunar disk as a function of the linear digits (with a maximum entry of 12) of the eclipsed lunar diameter. It is the same as Table TV 13. As was the case for solar eclipses, this table is already found in Ptolemy's *Almagest* (VI, 8), in the *Handy Tables* (Stahlman 1959, p. 258), as well as in a number of earlier sets of tables in use in medieval Spain. In chapter 7 of the *Ḥibbur*, Zacut mentions the origin of this specific table: "The second table is according to the opinion of R. Judah b. Asher and according to the opinion of Ptolemy" (MS B, f. 21b; cf. MS L, f. 206v, where part of the passage is missing; MS S, f. 23v: *la 2a tabla es a opinion de rabi yuda abenaser la qual tomo del ptholomeo*; cf. Cantera 1931, p. 180).

The Latin version of the *Almanach Perpetuum* also contains a list displaying the circumstances of solar and lunar eclipses from 1493 to 1525 (see Table AP 47).

AP 17. Table for the equation of syzygies

This table is intended to be used for calculating the time of true syzygy (Chabás and Goldstein 1997, pp. 97–98). Its heading in the *Almanach Perpetuum* is *Tabula ad verificandum horam aspectuum vel coniuntionis.*

This strange title seems to be a corrupt form of the heading in *ha-Ḥibbur ha-gadol*, for in MS L, f. 142r, it is "Table for correcting the time of conjunction and opposition, and quarters of the month and all astrological aspects of the Moon with all the planets" (see Table HG 14).

In this double argument table the vertical argument is the elongation (*arcus distantie*) between the Sun and the Moon, whose values are 0;5°, 0;10°, 0;20°, . . . , 1°, and thereafter for each half-degree to 13°, for a total of 31 values. The horizontal argument goes from 10;36° to 16° at intervals of 0;12°, and represents the daily increment in elongation. This table contains 868 entries, yielding the times in Salamanca, counted from mean noon, at which the true syzygies occur.

To recompute the entries, let λ_s and λ_m be the true longitudes of the Sun and the Moon respectively, at Salamanca for mean noon on the day when the syzygy takes place, and let λ_s^* and λ_m^* be the true longitudes exactly one day later. All these values are found in other tables of the *Almanach Perpetuum*, as previously noted: Tables AP 2 and AP 7 for the daily positions of the Sun and the Moon, respectively. Now, if the elongation, η, is $\lambda_s - \lambda_m$, and $\Delta\eta = (\lambda_m^* - \lambda_m) - (\lambda_s^* - \lambda_s)$ is the daily increment of elongation, we can recompute the entries, i.e., the time t counted from mean noon at which the true syzygy occurs, by means of the following equation:

$$t = 24 \cdot \eta / \Delta\eta. \qquad [8]$$

We have recomputed the 842 non-zero entries in this table using equation [8], above, and we obtained the following distribution (Z − C is the difference between the entry in Zacut's table and our recomputation):

Z − C	−2′	−1′	0′	+1′	+2′	≥ +3′
Frequency	11	75	597	144	10	5

If we allow for a precision of ±1′, the results given by Zacut are discordant in only 26 cases. However, the crucial issues do not concern accuracy, but rather: (1) why are we given this table to calculate the time of the true syzygies after we have been given a specific table (see Table AP 8, above) for these times, and (2) which of the two tables yields better results?

Chapter 5 of the *Ḥibbur* includes an explanation of this table, as well as the method for calculating true syzygies (Cantera 1931, pp. 170–173). There is even a worked example for the opposition of March 22, 1475. This seems to be the only method given by Zacut in his treatise for the computation of syzygies, for in the *Ḥibbur* there is no discussion of a table like Table AP 8, above. In the Castilian version of the *Almanach Perpetuum* (as well as in chapter 5 in the Latin version), there is an explanation of a method to compute syzygies. We are told that if we do not wish to use the table for syzygies computed by Jacob ben David Bonjorn (Poel), we can always use this table for the equation of syzygies which follows the opinion of King Alfonso (Albuquerque 1986, p. 82): *Aun que ayamos hecho tablas de coniunçiones et oposiçiones delos luminares proçedientes por aquella orden que proçedio el poel, enpero sy quisieres saber la coniunçion o oposiçion segund doctrina del serenissimo rrey don alfonso por estas tablas dela luna hechas por nos en esta manera se puede saber.*

The method described in the *Almanach Perpetuum* is similar to the explanation in the *Ḥibbur*. To check it against the method underlying Table AP 8, we have computed the time of the true conjunction on March 4, 1478; some of the magnitudes involved are:

$$\begin{aligned} \lambda_s &= 11\text{s } 23;12,21^\circ \text{ (Tables AP 2 and AP 4)} \\ \lambda_m &= 11\text{s } 20;29^\circ \quad \text{(Table AP 7)} \\ \eta &= 2;43^\circ \\ \Delta\eta &= 11;58^\circ \end{aligned}$$

Entering Table AP 17 with the last two values yields 5;27h (with interpolation). Adding 0;6h due to the equation of time (Table AP 5), we obtain the time, counted from noon at Salamanca, for the true conjunction: 5;33h. This time differs by almost half an hour from that found in Table AP 8 (6;0h), a table that always gives more accurate values. The reason is that the table we are examining here, computed by equation [8], above, is just a crude approximation of the complex problem of finding true syzygy.[70]

On the other hand, the reasons for associating this method with Alfonso X are not at all clear. It would seem that in the *Ḥibbur* Zacut proposed a relatively unsophisticated method for calculating true syzygies, and that some years later, he introduced a much more reliable procedure for finding the time of true syzygies.

AP 18. Table for lunar latitude

There are two tables for lunar latitude as a function of the argument of lunar latitude. Both of them have the heading, *Tabula latitudinis lune*. The entries in the first table are given to minutes for each integer degree of the argument; the maximum latitude (for argument 90°) is 5;0°. This table is very common in the medieval astronomical literature and it is found in Ptolemy, al-Battānī, Azarquiel, and the *editio princeps* of the Alfonsine Tables, among many others.

In the second table, which is intended for the computation of eclipses, the argument of latitude ranges from −6;0° to 17;0° and from 163° to 186°

[70] For an explanation of this problem and the methods proposed in the Middle Ages to solve it, see Chabás and Goldstein 1992.

at half-degree intervals. Here the latitude is given to seconds. Surprisingly, the entries are based on a maximum of 4;29°, rather than on 5;0° as in the first table. This parameter, 4;29°, is explicitly mentioned by Zacut in chapter 4 of his *Ḥibbur* (MS B, f. 15b; MS L, f. 210v; MS S, f. 14r: *as de saber que segun opinion de algunos de los postrimeros la ladeza maior de la luna no llega a mas de 4. grados e 29. minutos. e llamaron a esta ladeza la ladeza precisa*; cf. Cantera 1931, p. 168). Note that he does not cite any astronomer who had used it previously.

Very similar tables, with a maximum latitude of 4;30°, are found in al-Khwārizmī, Yaḥyā ibn Abī Manṣūr, and Levi ben Gerson.[71] But there are at least two Spanish zijes where we find a pair of tables for the lunar latitude, one with a maximum 5;0° and the other with 4;29°: *al-Muqtabis* of Ibn al-Kammād, and the Tables of Barcelona.[72] These sets of tables presumably belong to the direct line of recent predecessors (*postrimeros*) from whom Zacut took his tables for lunar latitude.

Table TV 10 in Madrid, MS 3385, is almost the same as Table AP 18, with the same range and basic parameter, but the argument is given at intervals of 0;15°, rather than 0;30°.

AP 19. Table for right ascension

This is a standard table for the normed right ascension, where the entries represent the right ascension, increased by 90°, as a function of longitude, beginning with Capricorn 0°. The entries are given in degrees and minutes, and were computed with an obliquity of 23;35°. This table is found, but for minor differences, in the zij of al-Battānī, the Toledan Tables, and in the Almanac of Azarquiel (in an abridged version), as well as in the 1483 edition of the Alfonsine Tables.[73]

Tables AP 20, AP 24, AP 28, AP 33, AP 38: the true longitudes of the planets

These five tables give the true longitudes of the five planets in degrees and minutes, and are calculated for the meridian of Salamanca for noon on various dates, beginning in March 1473.

---

[71] For the table in al-Khwārizmī's zij, see Suter 1914, pp. 132–134, and Neugebauer 1962, pp. 95–98. For a reference to Yaḥyā's zij, see Kennedy 1956, p. 146; and for the table in Levi ben Gerson, see Goldstein 1974, pp. 132–133 and 212–217, col. VI.

[72] For Ibn al-Kammād's zij, see Chabás and Goldstein 1994, pp. 20–22; for the Tables of Barcelona, see Chabás 1996a, p. 504.

[73] For this table in al-Battānī's zij, see Nallino 1899–1907, 2:61–64; for the Toledan Tables, see Toomer 1968, p. 34; for the abridged version of this table in the Almanac of Azarquiel, see Millás 1950, pp. 220–221.

For Saturn we are given its position on the ecliptic for days 10, 20, and the last day of each month for a period of 60 years. Note that in chapter 7 of the Latin version of the *Almanach Perpetuum* we are told that Saturn completes one revolution in 59 years, whereas the Castilian version erroneously has 56 years. In the case of Jupiter, positions are given for days 8, 16, 24, and the last day of each month for 85 years. The Latin and the Castilian versions agree in saying that the planet takes 83 years to complete a cycle. As for Mars, its true longitude is calculated for days 1, 6, 11, 16, 21, 26 of each month, and for a cycle of 80 years. According to the Latin and Castilian texts, Mars completes its cycle in 79 years. The daily positions of Venus are given for a cycle of 8 years, the time the planet takes to complete a cycle according to both versions of the *Almanach Perpetuum*. The true longitude of Mercury is given for days 4, 8, 12, 16, 20, 26, and the last day of each month in 125 years, and this is the number of years both versions associate with a cycle of Mercury. Except for Mercury, all other values for the goal-years of the planets agree with the medieval almanac tradition (the Almanac of Azarquiel, the Almanac of Jacob ben Makhir, and the Almanac of 1307).

In Madrid, MS 3385, and Segovia, MS 110, the corresponding tables for the positions of the planets both have the same epochs for each planet respectively: January 10, 1475 (Saturn), March 8, 1475 (Jupiter), March 1, 1475 (Mars), March 1, 1473 (Venus), and March 4, 1475, but differ from those in the *Almanach Perpetuum*. The planetary cycles in these two manuscripts do not always agree with each other or with those in the *Almanach Perpetuum*.

As was the case for the tables for the daily solar and lunar positions, the entries in the tables for planetary positions in the *Almanach Perpetuum* derive from the Alfonsine Tables as presented in the *editio princeps*. We have recomputed the first entry in each table, and excerpts of the calculations are given below. For the method of calculation, see, e.g., Poulle 1984, pp. 206–208. Once again, we have used 0;0,27,20d as the time difference in longitude between Salamanca and Toledo. All quantities below are expressed in sexagesimal form and given in degrees except for the date that is given in days since the Incarnation.

AP 20. Table for the true longitude of Saturn

Recomputation of the first entry in the table:

March 10, 1473 (noon)	2,29,21,57; 0,27,20
Mean motion	1,25;51, 9
Mean center	3,12;57,13
Mean anomaly	4,31;19,46
Equation of center	+1;32

Provisional eq. of anomaly	−6;11
Diversitas diametri	0;23
Equation of anomaly	−6;33
True longitude	1,20;50
Entry in the *Almanach Perpetuum*	Gem 20;49
Difference (Zacut − Recomputation)	−0; 1

AP 24. Table for the true longitude of Jupiter

Recomputation of the first entry in the table:

March 8, 1473 (noon)	2,29,21,57; 0,27,20
Mean motion	3,59;18,33
Mean center	1, 6;11,19
Mean anomaly	1,55;54, 5
Equation of center	−5;21
Provisional eq. of anomaly	+10;18
Diversitas diametri	0;29
Equation of anomaly	+10; 4
True longitude	4, 4; 2
Entry in the *Almanach Perpetuum*	Sgr 4; 1
Difference (Zacut − Recomputation)	−0; 1

AP 28. Table for the true longitude of Mars

Recomputation of the first entry in the table:

March 1, 1473 (noon)	2,29,21,48; 0,27,20
Mean motion	5,16;56, 7
Mean center	3, 2;13,41
Mean anomaly	0,31;22,32
Equation of center	+0;30
Provisional eq. of anomaly	+12;12
Diversitas diametri	0;48
Equation of anomaly	+13; 0
True longitude	5,30;26
Entry in the *Almanach Perpetuum*	Psc 0;26
Difference (Zacut − Recomputation)	0; 0

AP 33. Table for the true longitude of Venus

Recomputation of the first entry in the table:

March 1, 1473 (noon)	2,29,21,48; 0,27,20

Mean motion	5,48;18,40
Mean center	4,17;23, 4
Mean anomaly	1,14;49,24
Equation of center	+2; 8
Provisional eq. of anomaly	+29;30
Diversitas diametri	0;25
Equation of anomaly	+29;35
True longitude	0,20; 2
Entry in the *Almanach Perpetuum*	Ari 20; 2
Difference (Zacut – Recomputation)	0; 0

AP 38. Table for the true longitude of Mercury

Recomputation of the first entry in the table:

March 4, 1473 (noon)	2,29,21,51; 0,27,20
Mean motion	5,51;16, 5
Mean center	2,21; 6,19
Mean anomaly	2,33;26,35
Equation of center	−1;57
Provisional eq. of anomaly	+13;19
Diversitas diametri	1;36
Equation of anomaly	+14;45
True longitude	0, 4; 4
Entry in the *Almanach Perpetuum*	Ari 4; 2
Difference (Zacut – Recomputation)	−0; 2

Tables AP 21, AP 25, AP 30, AP 35, AP 40: the arguments of center of the planets

These five tables display the arguments of center, or simply the centers, of the five planets, in degrees and minutes. The entries are calculated for the meridian of Salamanca at noon of days 1, 11, and 21 of each month, beginning in January 1473, in contrast to the tables for the true longitude of the planets beginning in March 1473. We are given tables for 60 years for Saturn, 84 for Jupiter, 34 for Mars, and 4 for both Venus and Mercury. Note that these numbers do not agree with the periods considered for the longitudes of the planets. In Madrid, MS 3385, the corresponding tables for the center of the planets begin in January 1476 in all cases.

As we showed for many previous tables, the entries in the tables for the centers of the planets in the *Almanach Perpetuum* derive from the Alfonsine Tables as presented in the *editio princeps*. We have recomputed the first entry in each table (January 1, 1473), and excerpts of the calculations are given below. Once again, we have used 0;0,27,20d as the time difference in longitude between Salamanca and Toledo. All quantities below are expressed in sexagesimal form and given in degrees, except for the date, presented here in days since the Incarnation.

AP 21. Table for the center of Saturn

Recomputation of the first entry in the table:

January 1, 1473 (noon)	2,29,20,49; 0,27,20
Mean center	3,10;40,39
Equation of center	+1;16
True center	3,11;57
Entry in the *Almanach Perpetuum*	6s 11;57
Difference (Zacut – Recomputation)	0; 0

AP 25. Table for the center of Jupiter

Recomputation of the first entry in the table:

January 1, 1473 (noon)	2,29,20,49; 0,27,20
Mean center	1, 0;42, 9
Equation of center	−5; 4
True center	0,55;38
Entry in the *Almanach Perpetuum*	1s 25;38
Difference (Zacut – Recomputation)	0; 0

AP 30. Table for the center of Mars

Recomputation of the first entry in the table:

January 1, 1473 (noon)	2,29,20,49; 0,27,20
Mean center	2,31;18,34
Equation of center	−6; 2
True center	2,25;17
Entry in the *Almanach Perpetuum*	4s 25;18
Difference (Zacut – Recomputation)	−0; 1

AP 35. Table for the center of Venus

Recomputation of the first entry in the table:

January 1, 1473 (noon)	2,29,20,49; 0,27,20
Mean center	3,19;14, 0
Equation of center	+0;44

True center	3,19;58
Entry in the *Almanach Perpetuum*	6s 19;57
Difference (Zacut – Recomputation)	–0; 1

AP 40. Table for the center of Mercury

Recomputation of the first entry in the table:

January 1, 1473 (noon)	2,29,20,49; 0,27,20
Mean center	1,19;59,48
Equation of center	–2;54
True center	1,17; 6
Entry in the *Almanach Perpetuum*	2s 17; 6
Difference (Zacut – Recomputation)	0; 0

Tables AP 22, AP 26, AP 31, AP 36, AP 41: the anomalies of the planets

These five tables display the anomalies of the five planets in degrees and minutes. They are calculated for the meridian of Salamanca for noon of days 1, 11, and 21 of each month, beginning in January 1473, just like the tables for the centers of the planets, but in contrast to the tables for their true longitudes that begin in March 1473. We are given tables for 60 years for Saturn, 86 for Jupiter (note that there are 2 different columns for year 80), 35 for Mars, 9 for Venus, and 40 for Mercury. These values do not agree either with the cycles considered for the longitudes of the planets or with those for their centers. In Madrid, MS 3385, the corresponding tables for the anomalies of the planets begin in January 1476 in all cases.

Once again, the entries in the tables for the anomalies of the planets in the *Almanach Perpetuum* derive from the Alfonsine Tables as presented in the *editio princeps*. We have recomputed the first entry in each table (January 1, 1473), and excerpts of the calculations are given below. All quantities below are expressed in sexagesimal form and given in degrees except for the date that is given in days since the Incarnation; 0;0,27,20d is taken as the time difference in longitude between Salamanca and Toledo.

AP 22. Table for the anomaly of Saturn

Recomputation of the first entry in the table:

January 1, 1473 (noon)	2,29,20,49; 0,27,20
Mean anomaly	3,26;35, 0
True anomaly	3,25;19
Entry in the *Almanach Perpetuum*	6s 25;19
Difference (Zacut - Recomputation)	0; 0

AP 26. Table for the anomaly of Jupiter

Recomputation of the first entry in the table:

January 1, 1473 (noon)	2,29,20,49; 0,27,20
Mean anomaly	0,56;20,12
True anomaly	1, 1;24
Entry in the *Almanach Perpetuum*	2s 1;24
Difference (Zacut - Recomputation)	0; 0

AP 31. Table for the anomaly of Mars

Recomputation of the first entry in the table:

January 1, 1473 (noon)	2,29,20,49; 0,27,20
Mean anomaly	0, 4; 8,33
True anomaly	0,10;11
Entry in the *Almanach Perpetuum*	0s 10; 9
Difference (Zacut - Recomputation)	−0; 2

AP 36. Table for the anomaly of Venus

Recomputation of the first entry in the table:

January 1, 1473 (noon)	2,29,20,49; 0,27,20
Mean anomaly	0,38;26,56
True anomaly	0,37;43
Entry in the *Almanach Perpetuum*	Tau 7;42
Difference (Zacut - Recomputation)	−0; 1

AP 41. Table for the anomaly of Mercury

Recomputation of the first entry in the table:

January 1, 1473 (noon)	2,29,20,49; 0,27,20
Mean anomaly	5,20;49,39
True anomaly	5,23;44
Entry in the *Almanach Perpetuum*	10s 23;44
Difference (Zacut - Recomputation)	0; 0

Tables AP 23, AP 27, AP 32, AP 37, AP 43: the latitudes of the planets

These five tables for the latitudes of the planets all have the same structure, but this structure is different from that used by most other astronomers for the same purpose. In these double argument tables in the *Almanach*

Perpetuum, the vertical argument is the true anomaly, α, given in degrees, from 0s 0° to 6s 0°, at intervals of 6°, and the horizontal argument is the true center, κ, also given in degrees, at 6°-intervals for the full range of the center. The entries, $\beta(\kappa, \alpha)$, represent the latitudes of the planet, given in degrees and minutes.

In the tables, the extremal latitudes of the five planets are:

	Northern	**Southern**
Saturn	+3;2° $\kappa =$ 10s 8° to 10s 12° $\alpha =$ 6s 0°	−3;5° $\kappa =$ 4s 8° to 4s 12° $\alpha =$ 6s 0°
Jupiter	+2;5° $\kappa =$ 0s 14° to 0s 26° $\alpha =$ 6s 0°	−2;8° $\kappa =$ 6s 14° to 6s 26° $\alpha =$ 6s 0°
Mars	+4;21° $\kappa =$ 0s 0° $\alpha =$ 6s 0°	−7;30° $\kappa =$ 6s 0° $\alpha =$ 6s 0°
Venus	+7;22° $\kappa =$ 9s 0° $\alpha =$ 6s 0°	−7;22° $\kappa =$ 3s 0° $\alpha =$ 6s 0°
Mercury	+4;5° $\kappa =$ 3s 0° $\alpha =$ 6s 0°	−4;8° $\kappa =$ 8s 18° to 8s 24°, 9s 6° to 9s 12° (but −4;5° for $\kappa =$ 9s 0°) $\alpha =$ 6s 0°

In the Castilian version, the 10 sub-tables giving the latitude of Mercury are not in their proper order, for the sequence is: 1, 2, 5, 6, 3, 4, 7, 8, 9, and 10; whereas in the Latin version the sequence is: 7, 8, 9, 10, followed by 3 pages with other tables, after which we find sub-tables 1, 2, 3, 4, 5, and 6.

The only other double argument tables for planetary latitudes we have found are in the Oxford Tables of 1348 ascribed to Batecomb (see, e. g., Oxford, Bodleian Library, MS Rawlinson D 1227, ff. 64r-87r; cf. North 1977). As noted in chap. 2.2, there were two Hebrew versions of these Oxford Tables, and the one preserved in Munich, MS Heb. 343, includes the planetary latitude tables. Copies of these Oxford Tables in Latin are preserved in Cracow (Jagiellonian Library, MS 553, ff. 25r-65v, and MS 610, ff. 1v-42r; cf. Rosińska 1984, p. 322, item 1647), and in Vienna, where they are associated with John of Gmunden who died in 1442 (MS Vin. 2440, ff. 21r-74r, and MS Vin. 5151, especially ff. 131v-146r). Zacut

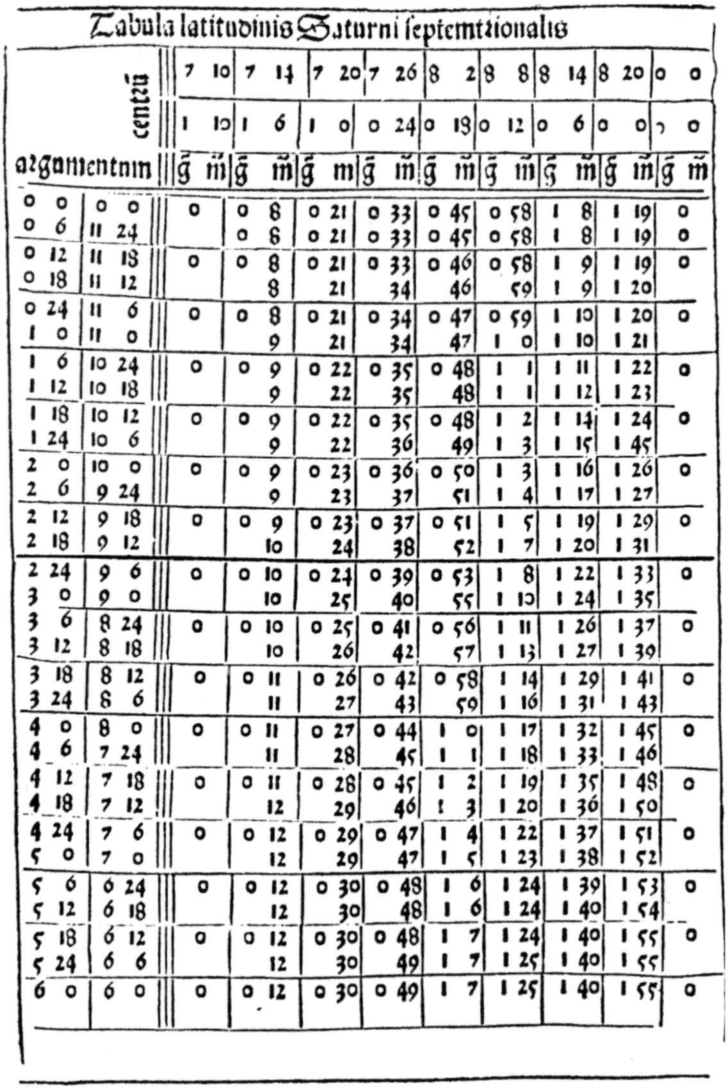

Tabula latitudinis Saturni septemtrionalis

argumentum		7 10 / 1 10	7 14 / 1 6	7 20 / 1 0	7 26 / 0 24	8 2 / 0 18	8 8 / 0 12	8 14 / 0 6	8 20 / 0 0	0 0 / 0 0
	centrum	g m	g m	g m	g m	g m	g m	g m	g m	g m
0 0	0 0	0	0 8	0 21	0 33	0 45	0 58	1 8	1 19	0
0 6	11 24		0 8	0 21	0 33	0 45	0 58	1 8	1 19	0
0 12	11 18	0	0 8	0 21	0 33	0 46	0 58	1 9	1 19	0
0 18	11 12		8	21	34	46	59	1 9	1 20	
0 24	11 6	0	0 8	0 21	0 34	0 47	0 59	1 10	1 20	0
1 0	11 0		9	21	34	47	1 0	1 10	1 21	
1 6	10 24	0	0 9	0 22	0 35	0 48	1 1	1 11	1 22	0
1 12	10 18		9	22	35	48	1 1	1 12	1 23	
1 18	10 12	0	0 9	0 22	0 35	0 48	1 2	1 14	1 24	0
1 24	10 6		9	22	36	49	1 3	1 15	1 45	
2 0	10 0	0	0 9	0 23	0 36	0 50	1 3	1 16	1 26	0
2 6	9 24		9	23	37	51	1 4	1 17	1 27	
2 12	9 18	0	0 9	0 23	0 37	0 51	1 5	1 19	1 29	0
2 18	9 12		10	24	38	52	1 7	1 20	1 31	
2 24	9 6	0	0 10	0 24	0 39	0 53	1 8	1 22	1 33	0
3 0	9 0		10	25	40	55	1 10	1 24	1 35	
3 6	8 24	0	0 10	0 25	0 41	0 56	1 11	1 26	1 37	0
3 12	8 18		10	26	42	57	1 13	1 27	1 39	
3 18	8 12	0	0 11	0 26	0 42	0 58	1 14	1 29	1 41	0
3 24	8 6		11	27	43	59	1 16	1 31	1 43	
4 0	8 0	0	0 11	0 27	0 44	1 0	1 17	1 32	1 45	0
4 6	7 24		11	28	45	1 1	1 18	1 33	1 46	
4 12	7 18	0	0 11	0 28	0 45	1 2	1 19	1 35	1 48	0
4 18	7 12		12	29	46	1 3	1 20	1 36	1 50	
4 24	7 6	0	0 12	0 29	0 47	1 4	1 22	1 37	1 51	0
5 0	7 0		12	29	47	1 5	1 23	1 38	1 52	
5 6	6 24	0	0 12	0 30	0 48	1 6	1 24	1 39	1 53	0
5 12	6 18		12	30	48	1 6	1 24	1 40	1 54	
5 18	6 12	0	0 12	0 30	0 48	1 7	1 24	1 40	1 55	0
5 24	6 6		12	30	49	1 7	1 25	1 40	1 55	
6 0	6 0	0	0 12	0 30	0 49	1 7	1 25	1 40	1 55	0

Figure 10. Table AP 23: Latitude of Saturn, first page only, *Almanach Perpetuum*, Leiria, 1496, f. 77v (Madrid, Biblioteca Nacional, I-1077).

did not simply copy the tables for planetary latitude in the Oxford Tables, for the arguments are defined differently and the entries generally differ as well. But, to be sure, the underlying models are the same.

For Mars, Venus, and Mercury, some of the extrema in Zacut's planetary latitude tables do not agree with those in the *Almagest* (XIII, 5). For Mars the extreme southern latitude in the *Almagest* is $-7;7°$, but the variant $-7;30°$ is attested in some copies of the Toledan Tables (Toomer 1968, p. 72) and in the *editio princeps* of the Alfonsine Tables (f. h1v). Similarly, the extrema for Venus are $\pm 7;22°$ according to Zacut, but $\pm 6;22°$ according to the *Almagest*. The variants $+7;24°$ and $-7;22°$ are found in some manuscripts of the Toledan Tables, but the *editio princeps* of the Alfonsine Tables has $\pm 7;12°$. In the Oxford Tables of 1348 the extreme southern latitude of Mars is $-7;7°$, and the extrema for Venus are $\pm 7;22°$. The only manuscript that we found with Zacut's variants for both Mars and Venus is Oxford, Bodleian Library, MS Can. Misc. 27 (f. 90v), in a set of tables by John of Lignères, and this codex is clearly associated with Salamanca (see chap. 2.2). In the case of the extreme southern latitude of Mercury, there seems to be a local disturbance in Table AP 43 for which we have no explanation. But in the Hebrew version of the Oxford Tables the corresponding value is also $-4;8°$ (Munich, MS Heb. 343, f. 166b), as it is in the Latin version (MS Vin. 2440, f. 68r, and MS Vin. 5151, f. 145r), without any such local disturbance.

The entries in the tables for planetary latitudes can be derived from the tables in the *Almagest*. Let c_i denote a column in the tables for the latitudes of the five planets in the *Almagest* (XIII, 5).

For the outer planets, $c_3(\alpha)$ and $c_4(\alpha)$ tabulate functions giving the northern or southern latitude of the planet with a true anomaly α. Moreover, $c_5(\omega)$ is an interpolation coefficient depending on the argument of latitude of the planet, ω, which is related to its center, κ, differently for each outer planet (Neugebauer 1975, pp. 216–221):

$$\begin{aligned}\omega &= \kappa + 50° \quad \text{(Saturn)}\\ &= \kappa - 20° \quad \text{(Jupiter)}\\ &= \kappa \qquad\quad\ \text{(Mars)}.\end{aligned}$$

The entries $\beta(\kappa, \alpha)$ in our tables can then be recomputed by the following formula:

$$\beta(\kappa, \alpha) = c_5(\omega) \cdot c_3(\alpha),$$

for nothern latitudes, i.e., when $270° \leq \omega \leq 360°$ and $0° \leq \omega \leq 90°$, or

$$\beta(\kappa, \alpha) = c_5(\omega) \cdot c_4(\alpha),$$

for southern latitudes, i.e., when $90° \leq \omega \leq 270°$. In particular, when $c_5 = 60'$, that is, when $\kappa = 310°, 130°$ (Saturn), $\kappa = 20°, 200°$ (Jupiter),

$\kappa = 0°$, 180° (Mars), we should find exactly the same entries as we find in the *Almagest*, in columns c_3 and c_4, respectively. This is indeed the case, except for the entries for Mars near $\kappa = 180°$, where we find an extremal southern latitude of −7;30° whereas the value in the *Almagest* is −7;7°.

For the inner planets, the rules are more complicated, and some peculiarities should be noted. For example, the argument for the interpolation coefficient $c_5(\omega)$ is related differently to the center, κ, than was the case for the outer planets (Neugebauer 1975, p. 223):

$$\begin{aligned} \omega &= \kappa + 90° \text{ (Venus)} \\ &= \kappa - 90° \text{ (Mercury).} \end{aligned}$$

Although, in the case of Venus the extremal value ±7;22° differs from that in the *Almagest* (±6;22°), the rest of the entries for κ = 3s 0° and κ = 9s 0° are the same as in the *Almagest*. As for Mercury, the extremal southern latitude in our table is −4;8° in the vicinity of κ = 9s 0°, but exactly −4;5° (the value in the *Almagest*) for κ = 9s 0°. The rest of the entries for that value of κ also agree with those in the *Almagest* (XIII, 5).

We conclude that the tables for the planetary latitudes in the *Almanach Perpetuum* were calculated according to the rules given by Ptolemy from tables ultimately deriving from the *Almagest*, but containing some variants for the extremal values of Mars, Venus, and Mercury.

The canons of the *Almanach Perpetuum* explain the origin of these tables for the latitudes of the planets. In chapter 20 of the Castilian version we read: *e esta ladeza de venus hezimos prinçipalmente segund la yntinçion de tolomeo en su almagesto e de alfragane e de albatogni e non segund la orden delas tablas de juan de lineros por que me paresçe mas verdadero la opinion de tolomeo*. Chapter 10 of the Latin version has a similar sentence, but adds the name of Averroes. However, in the canons of the *Ḥibbur* (chapter 14) we read: "To find the latitude of Venus . . . and we depended on the *Almagest* of Ptolemy and Ben Rushd [Averroes], and al-Farghānī" (MS B, f. 39a; missing in MS L; cf. MS S, ff. 45v-46r). Neither al-Battānī nor John of Lignères is cited here. But in a passage in MS B, ff. 39b-40a (not in MSS L, Mu, or S), John of Lignères is mentioned in connection with an observation made by Zacut on July 24, 1476, 8½ hours after noon, at which time the Moon occulted Venus (see Goldstein and Chabás 1999):

> Gloss (*haggaha*) concerning Venus: I saw, on Wednesday, July 24, 1476, 8½ hours after noon,[74] the Moon occult (*heḥeshikh*) Venus, and they

[74] On July 24, 1476, true noon at Salamanca (whose modern longitude is given as 5;40° W) took place at 12;28h UT for, according to P. Huber's program for planetary positions, the Sun culminated there at 12.47h UT (= 12;28h UT). Thus, given D. Herald's computation

were both close to the western horizon. When I saw this I was careful to check if [this occultation] agreed with the tables. The apparent conjunction in longitude with the equation of time and the lunar parallax in longitude produced exact agreement in longitude for the centers [of Venus and the Moon] at Vir 24;15°, and the Sun was at Leo 10½°. In [argument of] latitude the Moon was about 6° from the ascending node, and therefore its latitude was 0;31°. The lunar parallax was 0;56° to the south according to all its orbs, as follows (*keyṣad*: *lit.* how [is this computed]?): if [the Moon] had been at the apogee of its great orb as in conjunctions, then [its parallax] would be 0;48°, but because of its epicycle it was then 0;4° less; so the result is 0;44°. And on account of the great orb when the Moon is not at the apogee, add 0;12°; thus, the parallax was 0;56°. I then subtracted 0;31° to the north from 0;56° to the south, and the resulting apparent lunar latitude is 0;25° to the south. I also established (*tiqqanti*: *lit.* corrected) the latitude of Venus: I derived the corrected center as 39½° and its anomaly (*manah*) as 121;32°, and I derived the three [components of] latitude as I have described in [my] commentary to the tables and, according to the tables, the latitude is 1;21° to the north; but according to the tables of Lignères (Heb. *Liniris*) 1;50° to the north. According to what I derived from the tables, there was 1;46° between the center of Venus and the center of the Moon, where Venus was to the north; but according to Lignères (Heb. *Liniris*) 2;15°, where Venus was to the north. But according to what we saw, the Moon was more to the north when it occulted Venus. This is most surprising for, even if there were only 30 minutes between them, they would not even touch each other. All the more so, when [the distance between them] is 3½ times that amount. And even if we say that Venus also has a parallax, it would only be 0;6° as is explained in the *Epitome of the Almagest* [by Averroes].[75] Since I saw this difference [i.e., the difference between the computations and the observation?] with [my own] eye[s] contrary to the corrected latitude according to the tables, I wrote this down in a note [for myself]. Later

that for an observer in Salamanca the occultation began at 21;11h UT, it follows that this event occurred at 8;43h after local true noon (21;11 − 12;28 = 8;43). Since the equation of time on that day (according to Zacut's tables) is 12 min, and the expression, "8½ hours after noon," is to be understood as 8½ hours after mean noon, the time reported by Zacut is 8;42h after true noon, in very good agreement with modern recomputation. Note that for Zacut, as was the case for Levi ben Gerson, true noon is always prior to mean noon.

[75] The title given here, *Al-Magisti qaṣar*, seems to refer to Jacob Anaṭoli's Hebrew translation (1235) of Averroes's *Epitome of the Almagest* (*Qiṣṣur al-magisṭi*) that is not extant in the original Arabic. For the passage where the parallax of Venus is given as 6 minutes, see Lay 1991, iii:140 (in the chapter on the order of the planets). Averroes does not explain how he arrived at 6 minutes (corresponding to about 573 t.r. for the distance of Venus from the Earth). The parallax of Mercury and Venus is mentioned in Levi ben Gerson's *Astronomy*, chap. 133, where the value 18 minutes is given (without specifying whether it is for Venus or Mercury: Paris, MS Heb. 724, f. 252b:23; Paris MS Heb. 725, f. 223a:25).

I examined [the consequences of] this carefully for, if one were to say that the ascending nodes of Venus for the inclination (*neṭiʿa*) and the slant (*neliza*)[76] are always in the place where Ptolemy found them, it would indeed be true that the Moon occults Venus [at this time]. If so, it is appropriate to add 36° to the center of Venus, for the apogee is now at Cnc 1° whereas in Ptolemy's time it was at Tau 25°. In this way, they are corrected for finding the latitude of Venus according to the tables that I composed. Now we begin [the chapter] on Mercury [end of chap. 14].

This passage is extraordinary for several reasons. As noted above in chapter 1, there are very few observations in the works of Zacut and no other is described with this precision. Moreover, he compares his computation with that based on the tables of John of Lignères who is not mentioned anywhere else in the *Ḥibbur*. It is also a test of the theories for Venus and the Moon against an observation.

AP 29. Table for the correction of Mars

This table gives the correction to be added to the longitude of Mars after one cycle of 79 years, as a function of the daily velocity of the planet and its position at the end of the cycle. The velocity is given at intervals of 0;2°/d, and the sign in the heading of each column refers to 15° of that sign, as explained in the text, because the apogee of Mars is Leo 15°.

This table is mentioned in chapter 16 of the Castilian version and in chapter 9 of the Latin version, as well as in chapter 13 of the *Ḥibbur*, where there is a lengthier explanation.

Table AP 29A: Correction of Mars

Daily motion in °/d			**Correction in degrees**				
	Leo	**Cnc**	**Gem**	**Tau**	**Ari**	**Psc**	**Aqu**
0;39	1;33	1;37	1;40	1;44	1;48	1;52	1;55
...	...	...	...	...	...	...	...
0; 1	2;21	2;46	3;10	3;35	3; 0*	4;24	4;47
0; 0	2;24	2;48	3;13	3;37	4; 1	4;26	4;51
0; 1 *Retro*	2;26	2;51	3;16	3;41	4; 6	4;31	4;56
0; 3	2;29	2;54	3;20	3;46	4;12	4;38	5; 4
...	...	...	...	...	...	...	...
0;25	3; 4	3;33	4; 3	4;33	5; 3	5;33	6; 3

* *Sic*

[76] 'Inclination' and 'slant' are Neugebauer's terms for the components of the latitude for Venus.

AP 34. Table for the correction of Venus

This table gives the correction to be added to the anomaly of Venus after one or more cycles of 8 years have elapsed. It consists of 4 columns: columns 1 and 3 display the number of elapsed 8-year cycles (col. 1 ranges from 1 to 30, and col. 3 from 1 to 24), and columns 2 and 4 display the correction (the entries in col. 2 are given in days, hours, and minutes, whereas those in col. 4 are given in degrees and minutes). As expected, the entries in columns 2 and 4 are always in the same proportion: their ratio is the daily velocity in anomaly of Venus (0;36,59°/d).

This table is mentioned in chapter 20 of the Castilian version and in chapter 10 of the Latin version; chapter 14 of the *Ḥibbur* has a lengthier explanation. The Castilian text gives 1;27,30° for the correction to be added to the anomaly of Venus after one cycle, but the Latin text erroneously gives 1;37,30°. The value deduced from the entry for 24 cycles in col. 4 (35;1°) is 1;27,32,30°.

The same table, but limited to columns 1 and 2, is found in Segovia, MS 110, f. 60r (Table HG 50), with all the entries properly copied, in particular that for 23 cycles (54d 10;24h, whereas the *Almanach Perpetuum* reads 54d 10;22h).

An increment of 1;27,30° in 8 years implies a daily velocity in anomaly of Venus of about 0;36,59,27,21°/d, a value hardly different from that in the Alfonsine Tables (0;36,59,27,23°/d). It comes as no surprise that this table for the correction of the anomaly of Venus is based on the same Alfonsine parameter as was the table for the anomaly of Venus (Table AP 36).

AP 39. Table for the correction of Mercury

This table gives the correction to be added to the longitude of Mercury after one cycle of 125 years as a function of the daily velocity of the planet at the end of the cycle. The velocity is given at intervals of 0;8°/d. When

Table AP 39A: Correction of Mercury

Daily motion in °/d	Correction in degrees
1;55	0;33
. . .	. . .
0;59	40
. . .	. . .
0; 3	47
0; 5 *Retro*	48
0;13	49
. . .	. . .
1; 9	0;56

Mercury travels exactly at its rate of mean motion, its longitude is to be increased by 0;40° after 125 years. For each increment (decrement) of 0;8° in its daily velocity, the correction decreases (increases) by 0;1°.

Some examples of the use of this table are given in chapter 21 of the Castilian version and in chapter 11 of the Latin version, as well as in chapter 15 of the *Ḥibbur*.

AP 42. Table for the unequal motion of Mercury

For the use of this table see chapter 15 of the *Ḥibbur* (Cantera 1931, pp. 214–215). This is a double argument table for the daily progress of Mercury as a function of equated center and equated anomaly. The horizontal argument (equated center) is given at 6°-intervals from 0s to 6s; the vertical argument (equated anomaly) is given at 3°-intervals from 0s to 6s. The entries are given to minutes.

Zacut does not refer to any specific source, merely alluding to "those who made similar tables" (MS B, f. 41b; MS L, f. 195r; MS S, f. 48v; Cantera 1931, p. 215). We have only found one such candidate for this category: Judah ben Asher (Vatican, MS Heb. 384, ff. 372a-374a). His table uses signs of 60° (rather than 30° as in Zacut): both arguments are given at 6°-intervals, and the entries are given to seconds (rather than to minutes as in Zacut). Judah ben Asher also has similar double argument tables for the daily progress of the other planets (and the Moon) that are not found in any of the works of Zacut. This suggests that Zacut copied the table for Mercury from Judah ben Asher, or from a common source, and rounded the entries to minutes. Zacut's reason for excluding the other planets eludes us. Moreover, we have not succeeded in recovering the method or the parameters used to compute the entries in these tables.

AP 44. Table for sexagesimal multiplication

This is a double argument table for interpolation purposes. The horizontal argument ranges from 1 to 11, whereas the vertical argument ranges from 0;1 to 0;34 at intervals of 1 min; from 0;36 to 2;0 at intervals of 2 min; from 2;0 to 3;3 at intervals of 3 min; and from 3;3 to 3;11 at intervals of 2 min. The entries were computed by multiplying the corresponding arguments for the rows and columns.

AP 45. Fixed stars

The Leiria edition has a list of coordinates of 56 stars. It is indeed an odd list, for the names of the stars are omitted, making it of little value for any reader. For each star the entries are arranged as follows: magnitude ("1" or "2"), a blank column (presumably for the star names), longitude (signs,

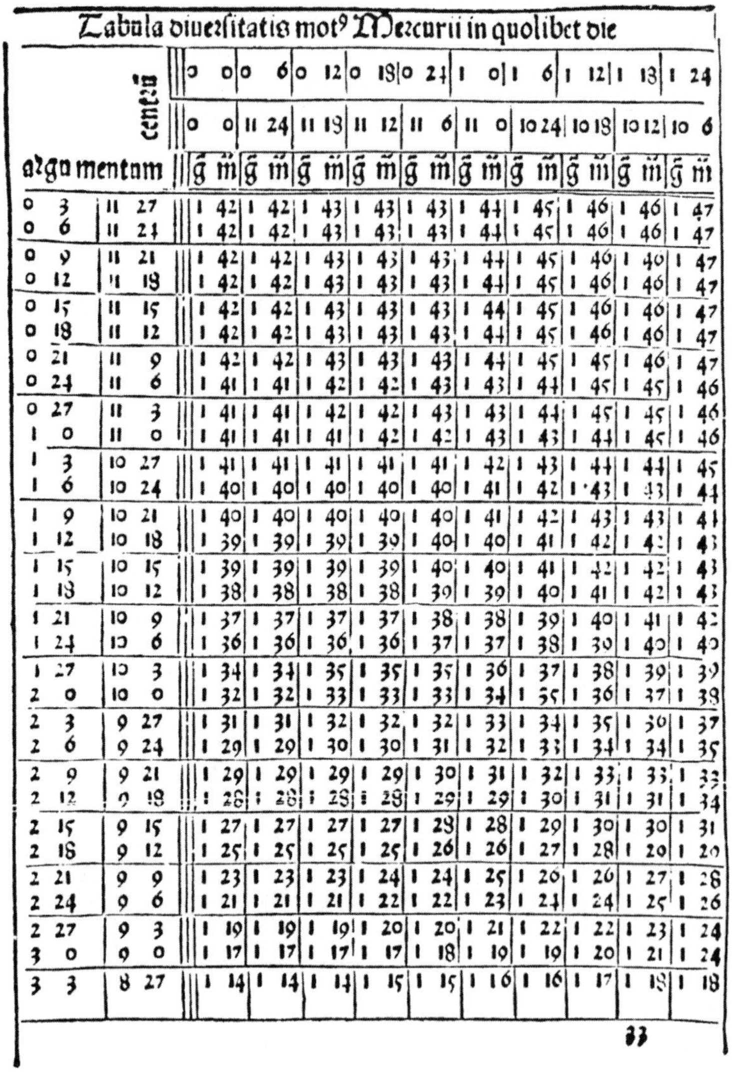

Tabula diuersitatis mot⁹ Mercurii in quolibet die

argumentum		g m	g m	g m	g m	g m	g m	g m	g m	g m	g m
centrum		0 0	0 6	0 12	0 18	0 24	1 0	1 6	1 12	1 18	1 24
		0 0	11 24	11 18	11 12	11 6	11 0	10 24	10 18	10 12	10 6
0 3	11 27	1 42	1 42	1 43	1 43	1 43	1 44	1 45	1 46	1 46	1 47
0 6	11 24	1 42	1 42	1 43	1 43	1 43	1 44	1 45	1 46	1 46	1 47
0 9	11 21	1 42	1 42	1 43	1 43	1 43	1 44	1 45	1 46	1 46	1 47
0 12	11 18	1 42	1 42	1 43	1 43	1 43	1 44	1 45	1 46	1 46	1 47
0 15	11 15	1 42	1 42	1 43	1 43	1 43	1 44	1 45	1 46	1 46	1 47
0 18	11 12	1 42	1 42	1 43	1 43	1 43	1 44	1 45	1 46	1 46	1 47
0 21	11 9	1 42	1 42	1 43	1 43	1 43	1 44	1 45	1 45	1 46	1 47
0 24	11 6	1 41	1 41	1 42	1 42	1 43	1 43	1 44	1 45	1 45	1 46
0 27	11 3	1 41	1 41	1 42	1 42	1 43	1 43	1 44	1 45	1 45	1 46
1 0	11 0	1 41	1 41	1 41	1 42	1 42	1 43	1 43	1 44	1 45	1 46
1 3	10 27	1 41	1 41	1 41	1 41	1 41	1 42	1 43	1 44	1 44	1 45
1 6	10 24	1 40	1 40	1 40	1 40	1 40	1 41	1 42	1 43	1 43	1 44
1 9	10 21	1 40	1 40	1 40	1 40	1 40	1 41	1 42	1 43	1 43	1 44
1 12	10 18	1 39	1 39	1 39	1 39	1 40	1 40	1 41	1 42	1 42	1 43
1 15	10 15	1 39	1 39	1 39	1 39	1 40	1 40	1 41	1 42	1 42	1 43
1 18	10 12	1 38	1 38	1 38	1 38	1 39	1 39	1 40	1 41	1 42	1 43
1 21	10 9	1 37	1 37	1 37	1 37	1 38	1 38	1 39	1 40	1 41	1 42
1 24	10 6	1 36	1 36	1 36	1 36	1 37	1 37	1 38	1 39	1 40	1 40
1 27	10 3	1 34	1 34	1 35	1 35	1 35	1 36	1 37	1 38	1 39	1 39
2 0	10 0	1 32	1 32	1 33	1 33	1 33	1 34	1 35	1 36	1 37	1 38
2 3	9 27	1 31	1 31	1 32	1 32	1 32	1 33	1 34	1 35	1 36	1 37
2 6	9 24	1 29	1 29	1 30	1 30	1 31	1 32	1 33	1 34	1 34	1 35
2 9	9 21	1 29	1 29	1 29	1 29	1 30	1 31	1 32	1 33	1 33	1 33
2 12	9 18	1 28	1 28	1 28	1 28	1 29	1 29	1 30	1 31	1 31	1 34
2 15	9 15	1 27	1 27	1 27	1 27	1 28	1 28	1 29	1 30	1 30	1 31
2 18	9 12	1 25	1 25	1 25	1 25	1 26	1 26	1 27	1 28	1 20	1 29
2 21	9 9	1 23	1 23	1 23	1 24	1 24	1 25	1 26	1 26	1 27	1 28
2 24	9 6	1 21	1 21	1 21	1 22	1 22	1 23	1 24	1 24	1 25	1 26
2 27	9 3	1 19	1 19	1 19	1 20	1 20	1 21	1 22	1 22	1 23	1 24
3 0	9 0	1 17	1 17	1 17	1 17	1 18	1 19	1 19	1 20	1 21	1 24
3 3	8 27	1 14	1 14	1 14	1 15	1 15	1 16	1 16	1 17	1 18	1 18

33

Figure 11. Table AP 42: Unequal motion of Mercury, first page only, *Almanach Perpetuum*, Leiria, 1496, f. 149r (Madrid, Biblioteca Nacional, I-1077).

degrees, and minutes), latitude (degrees and minutes, with the indication "south" [*meridi*] or "north" [*septem*]), and the planets associated with it for astrological purposes. The longitudes given in the *Almanach Perpetuum* differ by 6;38° from those for stars listed in Ptolemy's *Almagest* with the same latitude. This difference in longitude points to the Hijra as the epoch for the longitudes in this list.[77] This is also the case for Ibn al-Kammād's list of 30 stars (Goldstein and Chabás 1996), which is to be considered an ancestor of the present list.

The *Ḥibbur* contains a list (Table HG 33) that is almost identical with this one, but with 61 stars, including their names (see Table HG 33A).[78] As we noted in our article on Ibn al-Kammād's star list (1996, p. 327), the Leiria edition preserves the first 56 entries of the list in the *Ḥibbur* with very few variants for the coordinates and the "associated planets." For a discussion of the "associated planets," see Goldstein and Chabás 1996, pp. 319–324. In the Hebrew manuscripts there is a note concerning the last star in the list, α Centauri: "And for this star all the astronomers were in error, as Albohusain [al-Ṣūfī] taught, for they said that it is Libra; so wrote Judah ben Asher, and so I found it in the *Almagest*. But, in truth, it is in Scorpio, although we cannot decide since this star is not seen at our climate [i.e., geographical latitude]." In fact, the Arabic and Latin manuscripts of the star catalogue in the *Almagest* have α Cen in Libra, and al-Ṣūfī was aware of the difficulty, remarking that Scorpio is the correct reading as it is in the Greek manuscripts (Kunitzsch 1986–1990, 1:154–157; 2:158–159; Toomer 1984, p. 395).

A related list is found in a Latin version of the tables of the *Ḥibbur* (Madrid, MS 3385, f. 101v) with 83 entries, including 3rd magnitude stars and nebulae, but the longitudes listed in this manuscript are 14° greater than those for the same stars in the Leiria edition. For example, in the *Almanach Perpetuum* the longitude of an unnamed star is Leo 9;8°, and this is the longitude of Regulus (α Leo) in Ibn al-Kammād's list as well as in the *Ḥibbur* (MS L). But in Madrid, MS 3385, the longitude of Regulus is Leo 23;8°, i.e., 14° greater than in the *Almanach Perpetuum*. In the canons of the *Ḥibbur*, the longitude of Regulus in 1478 is given as Leo 23° with respect to the 9th sphere and Leo 9;8° with respect to the 8th sphere (Cantera 1931, pp. 298 ff). Zacut adds that in 1478 the difference between the 8th and the 9th spheres was 13;52°, and that the total precession since the time of Ptolemy's star catalogue is 20;30°. The difference between

[77] Other scholars believe that the epoch of this tables is January 25, 581, that is 41 years before the Hijra: see Comes 1991, p. 86; and Samsó 1994, p. 22.

[78] We wonder if the reason for the omission of the star names in the edition of 1496 was due to the ignorance of the translator who was not sure of the Latin equivalents for the Hebrew names.

Tabla stelaꝝ fixaꝝ p̄me ⁊ secūde magnitudinis ad ḡ octaue spere

magnitudo		signa	longit	lati	pars	natura
1		♈	6 48	61 30	meridi	Jovis
1		♉	6 18	23 0	septem	martis ⁊ mercu
1		♉	19 18	5 10	meridi	martis
1		♊	1 38	22 30	septem	Jovis ⁊ mercu
1		♊	8 38	17 30	meridi	martis ⁊ mercu
1		♉	26 28	31 30	meridi	saturni ⁊ Jovis
1		♊	23 48	75 40	meri i	saturni
1		♊	24 18	39 10	meridi	Jovis ⁊ mātis
1		♋	5 48	16 10	meridi	mercu ⁊ martis
1		♌	9 8	0 10	septem	Jovis ⁊ mātis
1		♍	1 8	11 50	septētr	saturni ⁊ veneris
1		♎	3 18	2 0	meridi	veneris ⁊ mercu
1		♎	3 38	31 30	septem	Jovis ⁊ mātis
1		♏	19 18	4 0	meridi	mātis ⁊ Jovis
2		♐	23 58	62 0	septem	veneris ⁊ mercu
1		♒	13 38	23 0	meridi	martis ⁊ mercu
2		♒	15 48	60 0	septem	veneris ⁊ mercu
2		♊	1 58	24 10	meridi	Jovis ⁊ saturni
2		♉	11 28	30 30	septem	martis ⁊ merco
2		♊	9 28	20 0	septem	martis ⁊ mercu
2		♊	0 38	17 30	meridi	Jovis ⁊ saturni
2		♋	24 48	72 10	septem	saturni ⁊ ueneris
2		♋	24 18	49 0	septem	martis
2		♋	28 48	44 10	septem	martis
2		♊	29 58	9 20	spetem	mercu
2		♋	3 23	6 15	septem	martis
2		♌	1 48	8 30	septem	saturni ⁊ mercurii
2		♌	2 48	74 50	septeui	saturni ⁊ uener

Figure 12a. Table AP 45: Fixed stars, *Almanach Perpetuum*, Leiria, 1496, f. 164v (Madrid, Biblioteca Nacional, I-1077).

Tabla stelarum fixarum secunde magnitudinis ad g octave spere					
magnitudo	signa	longit	lati	pars	natura
2	♌	9 18	47 30	septem	martis
2	♌	18 47	53 30	septem	martis
2	♌	2 48	13 40	septem	saturni ⁊ veneris
2	♌	6 38	20 30	meridi	martis ⁊ veneris
2	♍	6 38	54 0	septem	martis
2	♎	21 18	44 30	septem	veneris ⁊ mercu
2	♎	24 38	2 0	septem	saturni ⁊ mercu
2	♎	28 48	8 50	fepten	saturni ⁊ mercu
2	♏	4 38	4 45	septem	saturni ⁊ martis
2	♏	11 28	1 2	septem	saturni ⁊ martis
2	♐	23 58	23 0	meridi	iovis ⁊ mercu
2	♐	23 58	18 0	meridi	iovis ⁊ mercu
2	♑	10 28	29 10	septem	martis ⁊ iovis
2	♑	13 48	49 20	septem	veneris ⁊ mercu
2	♏	13 38	34 10	septem	saturni ⁊ martis
2	♓	25 58	26 0	septem	martis ⁊ veneris
2	♓	18 48	32 30	septeu	martis ⁊ mercu
2	♓	8 48	31 0	septem	martis ⁊ mercu
2	♓	2 18	19 40	septem	m̃tis ⁊ mercu
2	♉	3 18	40 30	sedtem	m̃tis ⁊ mercu
2	♉	8 58	3 40	septem	m̃tis ⁊ lune
2	♋	16 58	0 40	septem	m̃tis ⁊ lune
z	♐	7 48	13 30	m̃dio	m̃tis ⁊ lune
z	♐	21 48	0 45	septem	saturni et m̃is
2	♊	3 38	13 30	m̃dio	iovis
2	♈	13 8	7 28	septem	m̃tis et saturni
2	♋	23 8	5 30	m̃dio	saturni et m̃cur
2	♍	5 38	8 10	septeu	m̃rcur et m̃tis

Figure 12b. Table AP 45: Fixed stars, *Almanach Perpetuum*, Leiria, 1496, f. 165r (Madrid, Biblioteca Nacional, I-1077).

20;30° and 13;52° is 6;38°, which is the total precession between stellar longitudes in Ptolemy's catalogue and Ibn al-Kammād's list. Moreover, in an astrological work, Zacut gives the longitude of Regulus as Leo 23° (Carvalho 1927, p. 22). In the *Almagest* the longitude of Regulus is Leo 2;30° for the beginning of the reign of Antoninus (137 A.D.), and Zacut says that this date was 1341 years before his time (1478 A.D.), thus indicating that there was no confusion in his mind concerning the date of Ptolemy's catalogue.

The editions of 1502 and 1525 display the same table for 56 fixed stars that is found in the edition of 1496, without any star names. However, the 1502 edition has another list of 53 stars, for each of which we are given its name, longitude, latitude, declination, right ascension, magnitude, and associated planets (we have consulted the copy in Salamanca whose shelfmark is 38678). A facsimile of this list of 53 stars appears in Flórez *et al.* 1989, pp. 62–63, without specifying the edition. In general, the stellar longitudes in this list may be obtained by increasing the corresponding longitudes in Table AP 45 by 13°, while the latitudes are the same. In the preface to the 1502 edition, Alfonso de Córdoba explains (f. a2v) that he has adapted the star list for the year 1499 and has added equatorial coordinates. In this case, 13° is taken as the difference between the 8th and 9th spheres, rather than 14° (as in Madrid, MS 3385) that more closely agrees with Zacut's value of 13;52°. Evidently, Alfonso de Córdoba's computation of the equatorial coordinates required a considerable effort, and the star names in his list indicate that he used other sources in addition to Zacut, since the names are neither identical with those in Madrid, MS 3385, nor could they have been taken from the edition of 1496.

AP 46. Animodar

The last chapter of both the Castilian and the Latin versions of the *Almanach Perpetuum* is devoted to nativities. The Latin version has an explicit reference to sentence 51 of Ptolemy's *Centiloquium*, on the position of the Moon at the time of childbirth and the ascendant at the time of conception. The Castilian version mentions the book but not the sentence number.

The heading for the table is *De animodar ptholomei*, referring to the rules for nativities given by Ptolemy in his *Tetrabiblos* (III, 3). Al-Bīrūnī, in his *Book of Instruction in the Elements of the Art of Astrology*, gives an explanation of this term:

> Should no observation have been made at the time of birth, the determination of that time is beyond the reach of science, for there is no way of knowing it, but astrologers by estimation and conjecture arrive at one little different in the sign of the ascendant, when an attentive observer employs cautious questioning. But it is necessary that there should be a

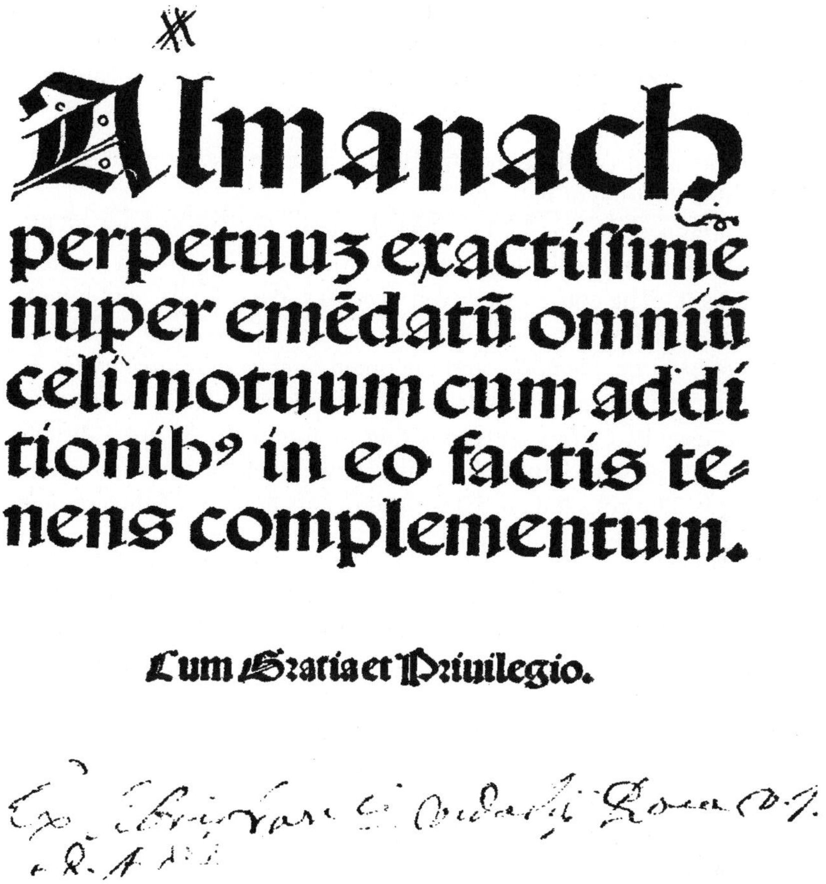
Almanach
perpetuuz exactissime
nuper emēdatū omniū
celi motuum cum addi
tionib9 in eo factis te
nens complementum.

Cum Gratia et Priuilegio.

Figure 13. Title page: *Almanach perpetuum exactissime nuper emendatum omnium celi motuum cum additionibus in eo factis tenens complementum.* (Venice: Petrus Liechtenstein, 1502).

> certain degree from the ascendant, so they find a way, by using an indicator [*namūdār*, whence "animodar" in the Latin translations] which furnishes one which they assume to be in the degree desired. The indicator most in use is that of Ptolemy . . . (Wright 1934, pp. 328–329).

In the medieval astrological literature there are several systems of *namūdār* to determine the time of nativity.[79]

The argument in this table is the *distantia lune* (argument of lunar latitude), and includes all integer degrees from 0s 1° (incorrectly written as 0s 0°) to 6s 0°. The entries are given in weeks, days, hours, and minutes, and they are displayed in two columns, one headed *Tempus more [infantis in utero matris] occidentalis*, and the other *Tempus more orientalis*. It is readily seen that consecutive entries differ by 1;49h or 1;50h in a sequence such that, usually, entries corresponding to arguments 3° apart differ by 5;28°, as in Table HG 68A. The maximum value in the first column (Western) is 39 weeks 0 days 5;15h, and occurs at 6s 0° of the argument; it seems to precede the minimum value in the second column (Eastern), 39 weeks 0 days 7;4h, which corresponds to an argument of 0s 1°. On the other hand, the entries in the two columns for any given argument differ by a constant (13d 16;0h).

This is not a common table in the astronomical literature, but we have located some others, and they may help in placing the table as printed in the *Almanach Perpetuum* in an astrological tradition. A table in London, Jews College, MS Heb. 135, ff. 166b-167b (a unique copy of the tables of Juan Gil of Burgos, *ca.* 1350) is virtually identical with that of Zacut, despite some differences in presentation and a few textual variants. Another table, very similar to Zacut's, is found in Madrid, MS 10023, ff. 62v-64r, under the title *Tabula extractionis quantitatis durationis creature in uentre matris per longitudinem lune a gradu occidentis*. It is described in a short canon in the same manuscript in Latin (f. 20r), which was transcribed by Millás, and is certainly a part of *al-Kawr ʿalā al-Dawr* by Ibn al-Kammād.[80] In the Madrid manuscript the entries in the column for the hours are blank, but one can definitely say that two consecutive entries, corresponding to arguments 1° apart, differ systematically by 1;50h, as opposed to the table in the *Almanach Perpetuum*. We have found the same table as that in Madrid, MS 10023 in several other MSS: Munich, MS Heb. 343, ff. 48a-49a (the related canon is on ff. 47a-47b); New York, Jewish Theological Seminary, MS Heb. 2597, ff. 67a-67b; Cracow, Jagiellonian Library, MS 609, ff. 82r-v. Still another table of this kind is in a manuscript in Arabic (of Maghribī origin): Escorial, MS Ar. 939 (ff. 5v-7r). This table

[79] For the various systems, see Kennedy 1995/96, pp. 139–144.
[80] Millás 1942, pp. 240–241; see also Chabás and Goldstein 1994.

was found by Vernet in a text of six chapters on astrological obstetrics, which he attributed to Ibn al-Kammād; Vernet (1949) published the table with a Catalan translation of that text. In fact, two different tables are given there: one is for a 9-month period of gestation, the other for a 7-month period of gestation. In both of them the entries are given for each integer degree of the argument. The entries in the first table differ by about 12h from those in the *Almanach Perpetuum*, but exhibit the same sequence of differences between consecutive entries. Although closely related, our table was not copied from any of the various tables described above (in Latin, Hebrew, and Arabic), and its relationship to them has not been established.

The table in the *Almanach Perpetuum* was reproduced some years later in a short set of tables computed for Leuven, Belgium (see chap. 5.4.2).

AP 47. Solar and lunar eclipses

On f. 167r of the Latin version there is a table, not included in the Castilian version, showing the circumstances of 6 solar eclipses and 16 lunar eclipses. For each of them we are given the date (day, month, and year), its magnitude (digits), the weekday, the time of the beginning, and the time of the end (both in hours and minutes). All the entries are shown in Tables AP 47A and 47B; we have added the Oppolzer number to identify each eclipse (Oppolzer 1887).

All these data seem to have been taken from the longer lists of eclipses in the *Ḥibbur* (see Table HG 31), where the lunar eclipses range from September 14, 1475 to August 25, 1523, and the solar eclipses from July 29, 1478 to June 7, 1518 or January 23, 1524 (in the year beginning in March); these two lists have been published in Chabás and Goldstein (1998), where full details are given. The lists of computed eclipses are mentioned in the canons of the *Almanach Perpetuum* (chapter 6 of the Latin version and chapter 8 of the Castilian version), where it is said: *Aunque fezimos eclipses ygualadas para 50 años*. The data for each eclipse in the *Almanach*

Table AP 47A: Solar Eclipses

Date	Mag.	Weekday	Beg.	End	Oppolzer No.
10 Oct. 1493	9	5	0; 0	1;20	6417
30 Sep. 1502	8	6	17;28	19;12	6438
20 Jul. 1506	3	2	1;49	3; 3	6446
7 Mar. 1513	4	1	23;49	1; 9	6462
7 Jun. 1518	10	2	18;22	19;17	6474
23 Jan. 1524*	9	2	3;12	4; 6	6489

* 1524 is found in Table HG 31, where the year begins in March; this date corresponds to 1525 in the year that begins in January.

Table AP 47B: Lunar Eclipses

Date	Mag.	Weekday	Beg.	End	Oppolzer No.
14 Sep. 1494	17	1	17; 5	2;33[a]	4176
18 Jan. 1497	17	4	3;50	7;18	4178
5 Nov. 1500	13	5	10;17	13;30	4182
2 May. 1501	19	1	15;33	19; 6	4184
15 Oct. 1502	14	7	10;15	12; 9	4186
29 Feb. 1504	16	5	10;47	14;13	4187
14 Aug. 1505	15	5	5;42	9; 6	4190
12 Jun. 1508	23	2	15;21	19; 0	4194
2 Jun. 1509	7	7	9;29	2; 3[b]	4196
6 Oct. 1511	13	2	9;11	2;25[c]	4199
29 Jan. 1514*	16	2	14;20	16; 3	4205
19 Jan. 1515**	15	7	5; 0	6;42	4207
13 Jul. 1516	14	1	10; 0	12;30	4208
6 Nov. 1519	20	1	5;50	6;48	4212
5 Sep. 1522	15	6	1;22	12; 4	4216
1 Mar. 1523	17	1	7;30	9;14	4217

* I. e., 1515 in the year that begins in January
** I. e., 1516 in the year that begins in January
a. Read 20;33?
b. Read 12; 3?
c. Read 12;25?

Perpetuum were taken directly, or deduced, from the *Ḥibbur* by the editor of the *Almanach Perpetuum*. Quite a few errors in transcription can be detected. It should also be noted that the time of the end of the eclipse (not listed in the *Ḥibbur*) is not always correctly reckoned from the time of its beginning and its half-duration (both listed in the *Ḥibbur*). For example, the entry for the end of the solar eclipse of June 7, 1518 (19;17h) results from adding to the time of the beginning (18;22h) the half-duration (0;55h), whereas twice the half-duration should have been added.

The 1502 edition of the *Almanach Perpetuum* displays the same table with the title *Tabula eclypsis luminarum in horizonte Salamanca*.

AP 48. Table for the half-length of daylight

This is another table found in the Latin (f. 167v), but not in the Castilian version. Its heading is *Tabula quantitatis dierum* and it is composed of 12 columns, where the first and last columns list the true solar longitude, λ, at intervals of 3°. The entries in the central columns correspond to geographical latitudes, ϕ, ranging from 36° to 45°, for each integer degree, and they represent the half-length of daylight (M/2), given in hours and minutes, for an observer at the appropriate geographical latitude and a given solar longitude. These entries can be calculated by means of the formula

$$M/2 = (90 + \arcsin(\tan\,\delta(\lambda) \cdot \tan\,\phi))/15, \qquad [9]$$

where $\delta(\lambda) = \arcsin(\sin\,\lambda \cdot \sin\varepsilon)$. The best results are obtained with $\varepsilon =$ 23;30°, a value for the obliquity of the ecliptic not attested in any other table in the *Almanach Perpetuum*.

We are persuaded that this table was simply taken from Regiomontanus's *Kalendarium* (Venice, 1483; printed by Erhard Ratdolt), for they both have exactly the same entries for the same arguments. But, in addition to the geographical latitudes in the *Almanach Perpetuum*, the table in the *Kalendarium* extends to latitudes up to 55°. All 310 entries in the table in the *Almanach Perpetuum* agree with those in the *Kalendarium*: even the errors agree! It is therefore unlikely that Zacut had anything to do with the inclusion of this table by Regiomontanus in the Latin version of the Leiria edition, just as we argued in the case of the dedication inserted at the beginning of that version.

AP 49. Table for geographical coordinates

This is the last of the three tables included in the Latin version of the *Almanach Perpetuum* (ff. 168r-169r), but not in the Castilian. It is a list of 92 cities, for each of which we are given its longitude and latitude (both in degrees and minutes). The list is not ordered by increasing longitudes or latitudes; it begins with Jerusalem (L = 67;30° and ϕ = 33;0°), and includes Salamanca (L = 25;46° and ϕ = 41;19°), Perpignan (L = 32;30° and ϕ = 42;30°), Toledo (L = 28;30° and ϕ = 39;54°), but not Leiria. More than half the cities named in it are in Spain.

This table clearly derives from Table HG 28 of the *Ḥibbur*, as shown by Cohn (1918, pp. 31–33) who published it together with the list in the *Ḥibbur* taken from MS Mu. Cantera (1931, pp. 363–370; and 1935, pp. 150–158) compared this list of coordinates with that in MS L (f. 133r), but gave no further comments. Recently, Laguarda has reconsidered Cantera's transcription, republished the table (twice!) and compared it with the entries in three Hebrew manuscripts of the *Ḥibbur* (Laguarda 1990, pp. 111–114 and 136–137).

AP 50. Calendar

This calendar just consists of over a hundred names of saints, virgins, popes, etc. associated with particular days of the year. It begins in January, and February has 28 days. It may be noted that Saint Adrian's day is taken here as March 3, whereas it is associated with March 4 in all the other fifteenth and sixteenth century calendars we have consulted.

AP 51. Sunday letters

The table at the bottom of the page yields the Sunday (or dominical) letter for each year in a 28-year cycle, beginning in 1473. For a given year, the Sunday letter yields the date in January of the first Sunday. For instance, the Sunday letter for 1473 is "C": this means that the first Sunday was January 3, and implies that year 1473 began on Friday (weekday 6). In the table at the top of the page we are given the number of the weekday at the beginning of each month of the year for a solar cycle of 28 years, beginning with 1473. For example, the number under January and Year 1 is 6.

AP 52. Movable feast days

This table displays the golden number (the number associated with each year in the 19-year lunar cycle) and the Sunday letter for a period of 35 years. For each year we are given the day of the movable feasts associated with Easter in the Latin Christian calendar. The 13 columns in this table fully agree with those in Regiomontanus's *Kalendarium* (f. c1v), as was the case in Table AP 48, above.

4.4 Dating the publication of the *Almanach Perpetuum*

The colophon of both the Castilian and the Latin versions reads:

> *Expliciunt tabule tabularum astronomice Raby abraham zacuti astronomi serenisimi Regis emanuel Rex portugalie et cet cum canonibus traductis alinga ebrayca in latinum per magistrum Joseph vizinum discipulum eius actoris opera et arte viri solertis magistri ortas curaque sua non mediocri impressione complete existunt felicibus astris anno a prima rerum etherearum circuitione 1496 sole existente in 15g 53m 35s piscium sub celo leyree.*

Fontoura da Costa (1983, p. 429) gave February 25, 1496 —with no further explanation— as the date for the printing of the *Almanach Perpetuum*. This is indeed the case, for this date corresponds precisely to the solar position mentioned in the colophon. According to Table AP 1, the position of the Sun for Feb. 25, 1475 (year beginning in March) is Psc 15;44,45°. This value has to be increased by the correction due to the 4-year cycles that elapsed from 1475 to 1496 that is displayed in Table AP 4: 0;8,50° for 5 cycles, or 20 years. The result is exactly Psc 15;53,35°, the value in the colophon. Note, however, that this value is valid for noon at Salamanca, but not for the "sky of Leiria," a city about 3° west of Salamanca.

4.5 Joseph Vizinus

According to the colophon, Joseph Vizinus was responsible for the preparation of the Leiria edition in 1496. In some copies of the Latin version a printed seal is found below the colophon, which bears the inscription: "IODE:VIZINHO." At the end of the canons of the Castilian version, for which there is no counterpart in the Latin version, we read:

> *Aqui se acaba la reçela delas tablas tresladadas de abrayco en latin e de latin en noestro vulgar romançe por mestre jusepe vezino deçipolo del actor delas tablas. deo graçias.*

As far as we know, these are the only contemporary documents where the surname Vizinus appears. Note that the spelling of "Vizinus" is different in each of the three cases, and that the modern convention has generally been to use the Portuguese spelling "Vizinho."

These documents raise several problems. The phrase in the colophon "translated from Hebrew to Latin" refers to the canons ("cum canonibus traductis"), but it can only apply to the headings of the tables, for there is no Hebrew counterpart to the canons in Latin (or Castilian). On the other hand, the explicit to the canons in Castilian seems to refer to the tables ("tablas tresladadas"), which were transcribed from Hebrew into Latin, but not from Latin into Castilian.

According to modern historians, Joseph Vizinus, the translator and editor of the *Almanach Perpetuum* in 1496, played an important role in Portuguese navigation during the previous decades.

Bensaude (1912, p. 105), following Schmeller, and ultimately based on a reference given by Valentim Fernandes (early sixteenth century), claimed that Joseph Vizinus was already a member of the entourage of King Afonso V of Portugal (reigned: 1438–1481) in about 1471: Valentim Fernandes reports that he had seen at the house of the Jew, "mestre Josepe," (no surname) a little statue offered to him as a present by the King.

Joseph Vizinus was supposedly a member of the so called "Junta dos Mathematicos" that gave advice to King João II of Portugal (reigned: 1481–1495) concerning navigation in 1482. As we noted in chapter 1, the literature on this "Junta" is indeed confusing and not very convincing. For the historian and chronicler João de Barros (ca. 1496–1570), and author of an apologetical historical work in four parts, *Décadas da Ásia*, this group of advisers included "D. Diogo Ortiz Bispo de Cepta, e com Mestre Rodrigo, e Mestre Josepe."[81] (Note again that no surname is given for "Mestre Josepe.") However, the expression, "Junta dos Mathematicos," has also been used to designate a different group of royal counselors appearing later in de Barros's account, when he explains that King João proposed to "Mestre Rodrigo, e a Mestre Josepe Judeo, ambos seus medicos, e a hum Martim de Boemia" to address the problem of navigation at sea by

---

[81] *Década I*, book 3, chap. 11, quoted as document 10 by Bensaude (1912, pp. 263–264).

means of the solar altitude. (Note that the surname is still lacking.)[82] The expression "Junta dos Mathematicos" of which this Joseph is supposed to have been a member seems to be a modern expression that is not found in de Barros (at least it is not found in the passages quoted by Bensaude).[83] Moreover, we find the scanty evidence available insufficent to conclude that Joseph Vizinus, the translator and editor of the *Almanach Perpetuum*, is to be identified either with the "Joseph" mentioned by Fernandes, or a "Joseph" mentioned by de Barros (and the two occurrences of "Joseph" in the text of de Barros need not refer to the same person).

Joseph Vizinus is also said to have been sent on a mission to Guinea in 1485 by the King of Portugal, to measure the altitude of the Sun in order to determine its geographical latitude. The documents mention no Vizinus, but a certain "magister Josepius fixicus et astrologus" (i.e., physician and astrologer). These documents are, in fact, brief marginal notes in two books owned by Christopher Columbus: the *Ymago mundi* by Pierre d'Ailly (Leuven, ca. 1483), and the *Historia rerum ubique gestarum* by Aeneas Silvius Piccolomini (Venice, 1477).[84] In the second document, the annotator —supposedly Columbus himself—asserts that he was present when the report was given to the King on the voyage of "magister Josepius" to Guinea, where he was sent in 1485 and where he made observations on March 11. We see several problems with these passages, e.g., we are told that Columbus had been living in Portugal since 1476, and secretly departed from Portugal to Castile at the beginning of 1485: this would mean that he left Portugal before the report was submitted to the king.[85]

From this story of the voyage by "magister Josepius" to Guinea, several modern historians of Portuguese discoveries have considered Vizinus as the author of the nautical part of the so called "Regimento de Munich," a

---

[82] *Década I*, book 4, chap. 2, quoted as document 8 by Bensaude (1912, pp. 261–262). For an English translation of this passage see Maddison (1992, pp. 97–98), who adds "José Vizinho," in square brackets, after "Josepe Judeu." For a Castilian translation of this passage see Albuquerque (1991, p. 156), who also adds "José Vizinho," in parentheses, after "José Judío." Note also that Albuquerque (p. 139) adds the phrase "both Jews" after the names of Joseph Vizinus and master Diogo (instead of master Rodrigo, as in the document) when referring to the contradictions between the two chroniclers, João de Barros and Gaspar Correia. "Martim de Boemia" is the German cosmographer Martin Behaim, born in Nuremberg in 1459 and died in Lisbon in 1507.

[83] Fontoura (1983, p. 14) is one of several modern Portuguese historians of navigation who does not accept this designation, and he refers to the "injustamente denominada Junta dos Matemáticos de D. João II."

[84] See Bensaude (1912, pp. 107–108), and Fontoura (1983, p. 37). The marginal notes in these books are very numerous (almost 1,000), and need to be analyzed carefully, for it has been argued that some of them may have been written by his brother Bartholomew, and even by Christopher's son, Ferdinand, in addition to Columbus himself.

[85] The literature on Columbus is vast, but see especially Arranz 1985, pp. 55–56.

set of directions to help pilots and sailors navigate in the open ocean.[86] We consider this claim to be mere speculation with no documentary evidence to support it, and we leave further discussion to specialists in the history of navigation.

Of course, we do not wish to deny the contributions made by "mestre Josepe"; rather, our purpose is only to cast some doubts on the attribution of all this activity, sometimes based on very weak evidence, to the same person (let alone identifying that person with Joseph Vizinus), for the name is common and the surname is not given. On the other hand, Joseph Vizinus, the translator and editor of the *Almanach Perpetuum*, is called the disciple of Zacut only in the colophon of that book. This has often been used as an argument to strengthen Vizinus's authority on astronomical matters. We believe it is difficult to reconstruct the relationship of Vizinus and Zacut on the basis of the documents available to us, but we note that Vizinus is not mentioned anywhere by Zacut, as far as we can determine.

According to Roth (1936–1937 and 1937–1938), Joseph Vizinus was forcibly converted to Christianity in 1497 when the practice of Judaism was declared illegal in Portugal, and had a son named Ezra who settled in Ferrara and wrote a text on the Jewish calendar. Ezra Vizino was certainly a member of the Jewish community and refers to his father with pious epithets in Hebrew that would not be appropriate for an apostate. So, it may be the case that his father returned to Judaism at some point, or that Ezra did not consider his father's conversion to be legitimate.

Ezra Vizino's text begins (Oxford, Bodleian Library, Opp. Add. 12° 149, f. 274a):

> This calendar is a gift sent to the master, the honorable (*naʿaleh*) R. (*k[evod] h[a-]r[av]*) Joseph Sasson, may his light continue to shine (i.e., may he continue to live for a long time: *y[a'ir] n[er]o*), by the master, the honorable, the venerable, calendar-maker (*ha-tokhen*), R. Ezra Vizino, may his light continue to shine. Tables of the festivals, *parshiyot* [weekly portions of the Pentateuch], *hafṭarot* [weekly portions from the Prophets], and the mean new moon (*molad shaweh*) were arranged by me, Ezra Vizino, the son of the great and famous calendar-maker (*ha-tokhen*), R. Joseph Vizino, may the memory of the righteous be for a blessing (*z[ekher] ṣ[addiq] l[ivrakha]*) . . . [for a cycle of] 247 years from 5340 to 5586 [= 1579–80 to 1825–26].

Note that the Hebrew term, *tokhen*, probably means "calendar-maker" here, rather than "astronomer" (or "astrologer") which is also possible.

The calendar of Ezra Vizino is bound with a prayer-book (*Temunot Teḥinnot Tefillot Sefarad*) in the copy in Oxford. This prayer-book has a

[86] Document 1 presented by Bensaude (1912, pp. 217–231).

colophon on f. 272a indicating that the printing was completed on Sivan 29, 5344 A.M. (= June 8, 1584) in Venice at the printing house of Juan de Gara at the command of Joseph ben Jacob Sasson. The title page of Ezra Vizino's calendar appears on f. 273a; it has no date or place of publication, and it ends on f. 287b.

5. The influence of Zacut's astronomical works

5.1 Disciples

In the sources there are only two people who are called disciples of Zacut: Joseph Vizinus (in the colophon of the *Almanach Perpetuum*), and Augustinus Ricius (in his only known astronomical work, *De motu octavae Sphaerae*, 1st edn. Trino, Italy, 1513, and 2nd edn. Paris, 1521). In chap. 4.5 we have discussed Vizinus and his relationship to Zacut; here we will consider what Ricius tells us about himself and about Zacut. It is evident that Ricius knew Hebrew, for he cites various works in Hebrew, translating their titles into Latin. For example, he refers to Zacut's treatise as "magna editione" (edn. 1521, ff. 6v, 51r) or as "magna compositione" (ff. 32r, 33r), both of which are translations of the Hebrew title, *ha-Ḥibbur ha-gadol.* Moreover, he cites Levi ben Gerson's book as "Milhamot Hascem, hoc est defensionum dei," that is, first he transliterates the Hebrew title and then translates it (f. 11r). Ricius mentions many Jewish scholars and classical authors, most of whom were not mentioned by Zacut. He calls himself a disciple of Zacut (*Abraham Zacuth, quem praeceptorem in astronomia habuimus* . . . , f. 29r), and indicates that he studied with him in North Africa (*Habraham Zacuth, astronomiae nostra tempestate peritissimus* . . . , *nobisque eum legentem in Africa apud Carthaginem audientibus* . . . , f. 6v). In his treatise, Ricius mentions Zacut a total of 8 times as one of the authorities for establishing the motion of the eighth sphere, and refers to Zacut's observation of an occultation of Spica by the Moon in 1474, reported in the *Ḥibbur* but not mentioned in the *Almanach Perpetuum.*

5.2 Later editions of the *Almanach Perpetuum*

All, or some of, the tables in the *Almanach Perpetuum* were reprinted several times in astronomical texts published in the late fifteenth and sixteenth centuries, specifically in 1498, 1502, 1525, and 1528, indicating that this book had become part of the astronomical tradition. All four editions were printed in Venice, in three different printing houses. It has also been claimed that there were other incunabula editions of the *Almanach Perpetuum*, but no references for them are given, and no extant copies have been identified.

The book most closely related to the *Almanach Perpetuum* was printed by Johannes Lucilius Santritter in Venice (1498) under the title *Ephemerides sive Almanach Perpetuum*, just 2 years after the Leiria edition. The tables are preceded by a short text of six pages explaining that the tables were computed for noon of the last day of February 1472 (completed year) for the meridian of Toledo (*sic*, instead of Salamanca). Neither Zacut's name nor that of Vizinus is mentioned anywhere in the book, but Johannes Regiomontanus is cited several times. This short text is preceded by a calendar and two tables (for geographical coordinates and the equation of time) derived from Regiomontanus's *Kalendarium* that was also printed in Venice (1483) by Santritter. For these reasons this edition of 1498 has often been attributed to Regiomontanus.

Again in Venice, in 1502 Petrus Liechtenstein printed a book entitled *Almanach perpetuum exactissime nuper emendatum omnium celi motuum cum additionibus in eo factis tenens complementum*. It is in fact a new edition of the *Almanach Perpetuum* printed in Leiria in 1496. The Venice edition of 1502 has annotations and corrections by Alfonso de Córdoba, a physician in the service of Cardinal Borgia in Rome. Alfonso de Córdoba was born in Seville, probably in 1458, and is probably to be identified with "Hispalensis" (i.e., a man from Seville) in Copernicus's *Commentariolus* (but see Swerdlow 1973, pp. 451–452, for literature on opposing views). He is the author of a small treatise on an instrument (*Lumen coeli seu expositio instrumenti astronomici a se excogitati,* printed in Rome, 1498), and a set of astronomical tables, based on the Alfonsine Tables (*Tabule Astronomice Elisabeth Regine*, first printed in Venice, 1503). Alfonso de Córdoba's edition of the *Almanach Perpetuum* does not mention Zacut's name in the title, but it does appear a few times in the introduction; the name of Vizinus is also mentioned there once. After the introduction we find the same dedication as in the Leiria edition (see chap. 4.1), here entitled *Epistola auctoris ad episcopum Salamantice*, and then there is a text organized in "problems" following the Latin canons of the Leiria edition, sometimes *verbatim*. Alfonso de Córdoba added a few tables in his edition that are not in the edition of 1496, and among them is a list of stars (see comments to Table AP 45 in chap. 4.3).

Lucas Antonius Iunta, another printer in Venice, was responsible for the next edition of the *Almanach Perpetuum* (1525). This time Zacut's name appears in the title, as well as those of the editors: *Almanach perpetuum sive tacuinus. Ephemerides et diarium Abraami zacuti hebrei. Theoremata aut. Joannis Michaelis germani buduren. Cum L. Gaurici doctoris egregij castigationibus et plerisque tabellis nuper adiectis*. It is noteworthy that this edition also contains the dedication to the bishop, as was the case in the Leiria edition, but the title of it has increased by one word, that of Zacut: *Epistola auctoris çacuti ad episcopum Salamantice*, thus attributing to Zacut a dedication for which he was not responsible. The canons follow very closely those in the 1502 edition. Note also that the only star list in this edition includes no star names, as was the case in some of the previous editions of the *Almanach Perpetuum*.

Finally, in Venice Petrus Liechtenstein reprinted in 1528 the edition of 1502 annotated by Alfonso de Córdoba with no further changes.

5.3 Zacut's impact on the Jewish community

After leaving Portugal in about 1497, Zacut lived in North Africa, and then migrated to Jerusalem. In North Africa he composed a set of tables with radix 1501, 28 years (i.e., a solar cycle) after 1473, and this set only survives in a few fragmentary texts. Later, in Jerusalem, he composed another set of tables with radix 1513, this time arranged for the Jewish calendar, and it too only survives in fragmentary texts. The tables for 1501 were the subject of a Hebrew commentary by Abraham Gascon in Cairo who computed a solar eclipse for 1542. The parallax table for Jerusalem included in the 1513 tables was cited in an astronomical work by Ḥayyim Vital (1542–1620), best known as a Jewish mystic. One partial copy of the 1513 tables has marginal notes by Jacob Mizraḥi of Aleppo, dated 1685, and Simon ben Jonah Mizraḥi composed a set of astronomical tables in Baghdad in 1596 in which he mentions Zacut, as well as his own brother, Jacob, "who followed the path of Abraham Zacut" (Goldstein 1981).

In addition to the four Hebrew manuscripts of the *Ḥibbur* that we have used, there are several later copies: New York, Jewish Theological Seminary (JTSA), MS Heb. 2602, that includes Zacut's canons copied in the seventeenth century from MS B; New York JTSA, MS Heb. 2554, copied in the seventeenth century; and New York, JTSA, MS Heb. 8112, copied in the eighteenth century, with additional notes in Spanish written in Hebrew characters. Also, it had previously been thought that New York, JTSA, MS Heb. 296 (= Benaim MS 44), a seventeenth century text from Morocco, was another copy of Zacut's tables, but this is only partly true, for most of it comes from the almanac of 1307 (Goldstein 1981, p. 247; Chabás 1996b, p. 264). Some additional material related to Zacut is also found in Jerusalem, Benayahu MS 136 (Jerusalem, National Library, Institute for Microfilmed Hebrew Manuscripts, Film No. 44753), that dates from the seventeenth or eighteenth century.

An Arabic version of the *Almanach Perpetuum* was composed by Moses Galiano in 1506–1507 in Istanbul, and it is extant in Escorial, MS Ar. 966 (Vernet 1979, pp. 333–351; and Derenbourg 1941, p. 111). The translator, also known as Moses ben Elijah Galina, was active in the Jewish community (Seligsohn 1904, 5:554–555), and so it is likely that this version was intended for a Jewish audience. From the numerous passages of the Arabic text translated by Vernet, it follows that Galiano's canons are not a translation of either the *Almanach Perpetuum* or of the *Ḥibbur*, but an adaptation of the *Almanach Perpetuum*. Note that the end of Galiano's canons concerns the *animodar*: this topic is discussed in the last chapter of the *Almanach Perpetuum*, but it is not included in the *Ḥibbur*. Moreover, Galiano has a worked example for a lunar eclipse of 1505 that is found neither in the *Almanach Perpetuum* nor in the *Ḥibbur*.

A Ladino version of the *Almanach Perpetuum*, was published in Salonica in 1568 by Daniel ben Peraḥia and, according to Cantera (1935, pp. 68–69), it has 23 chapters, ending with the chapter on the *animodar*, as in the Castilian version, printed in 1496. New York, JTSA, MS Heb. 5782, contains a copy of this Ladino version.

5.4 Zacut's impact on Christian scholars

Further evidence for the fame of Zacut and his tables is found in many authors of this period, particularly (but not exclusively) among scholars active in Salamanca.

5.4.1 The impact of the *Ḥibbur*

Segovia, MS Catedral 110, contains a partial set of tables of the *Ḥibbur*.[1] After the tables for the entry of the Sun into the zodiacal signs and the lists of lunar and solar eclipses, and preceding those for the daily solar positions, there is a short text in Latin (ff. 9r-11v). It consists of 8 unnumbered brief chapters, or notes, all written in the same hand as the rest of the manuscript, except for a worthless table on ff. 49v-52r.[2]

The first chapter (ff. 9r-v) explains the use of Table HG 9 (= Table AP 7), which gives the daily positions of the Moon for a period of 31 years beginning in March 1473. On the basis of the explanations given in the text, it is clear that this chapter follows more closely chapter 3 of the *Ḥibbur* than chapter 4 of the *Almanach Perpetuum*. For instance, we are told that the correction to be applied to the lunar longitude after a cycle of 31 years varies from 0;33° to 0;43° (an error for 0;49°); this information is explicitly stated in the *Ḥibbur* but not in the *Almanach Perpetuum*.

The second chapter (f. 9v) is a paragraph of 14 lines, and it is a paraphrase of a short part of chapter 1 of the *Ḥibbur*. It concerns solar declination (Table HG 3, identical to Table AP 3), a subject that is not considered in the canons of the *Almanach Perpetuum*.

---

[1] For a description of this manuscript and an analysis of some of its tables, see Chabás and Goldstein 1998.

[2] For each day of the year this table gives 2 entries, one in degrees, and the other in hours and minutes. The entries in degrees increase systematically by 1°/d, and those in units of time increase systematically by 0;4h/d, thus making this table a very crude approximation of any real astronomical motion. On the other hand, if we disregard the column for the argument, this table gives the correspondence between time degrees and hours, where 360° = 24h.

The third chapter (f. 10r-v) deals with the determination of the position of Venus and, in particular, mentions the table for the correction of Venus (Table HG 50, corresponding to Table AP 34). The instructions indicate that the epoch is 1473, as is the case in the corresponding table for the longitude of Venus in this manuscript (ff. 59r-63r, corresponding to Table HG 49).

The fourth to seventh chapters deal respectively with the positions of Mercury, Saturn, Jupiter, and Mars, as well as with the epochs and revolutions displayed in the tables for each of these planets. The paragraph concerning Mercury (ff. 10v-11r) mentions Profatius and his errors, whereas that for Jupiter (f. 11r) mentions the Alfonsine Tables. The last paragraph (f. 11v) contains notes for domification and judgments.

While it is clear that these notes are not a translation of any chapter of the *Ḥibbur*, it is of some interest that they contain a passage where the author writes in the first person, and refers to himself as the author of an almanac that shares some special features with the *Ḥibbur*: "Mercurius quidem habet radicem in nostro almanach in anno 1475 in marcio et fecimus eum equari de 4or in 4or dies pro 20 annis . . . Sed post lapsum istorum 20 annorum quos posuimus pro una revolutione . . ." (f. 10v). Despite the clues given here, this passage does not furnish enough evidence to identify the author of these notes, or to establish his relationship to Zacut.

Diego de Torres is known to have been active at the University of Salamanca beginning in 1469. During the following 10 years he held various positions until he was nominated for the chair of astronomy in the 1480s. The date cannot be given more precisely, but there is evidence that he was professor of astronomy in 1485 and 1487. Moreover, until 1480 Fernando de Fontiveros occupied the chair of astronomy.

Diego de Torres was the author of two works. First, a short treatise, entitled: *Eclipse del Sol. Medicinas preservativas y curativas y remedios contra la pestilencia que significa el eclipse del sol del año de mill & cccc.lxxxv a xvi. de marҫo*, published in Salamanca in 1485 (Amasuno 1972). The data for this solar eclipse, that took place on March 16, 1485, seem to derive from the table of eclipses in Zacut's *Ḥibbur*, in which case it is the earliest use of a table of the *Ḥibbur* by a Christian scholar, and well before the publication of the *Almanach Perpetuum* in 1496.

The second work by Diego de Torres, *Obra astrológica*, was written in Castilian, although its introduction and explicit are in Latin; the only known copy is: Madrid, Biblioteca Nacional, MS 3385, ff. 154r-185r. The explicit indicates that the author is "didacum de torres," master of arts and medicine, and professor of astronomy at the University of Salamanca. It also gives the date when the treatise was finished: May 25, 1487 (f. 185r). This work is divided into 4 parts (Cantera 1931, pp. 137–139), and the first part is entirely devoted to mathematical astrology. In the last chapter of the first part (*Para saber como asentaras los planetas en la figura despues que tovieres asentados los signos en ella*: chapter 9), the following planetary positions, for an unspecified date, are given as an example:

Sun	Ari	0°
Mercury	Ari	15°
Venus	Psc	26°
Mars	Ari	24°
Lunar node	Cnc	28°
Moon	Sco	16°

This position of the Sun is its entry into Aries (i. e., vernal equinox), and the planetary positions agree with those corresponding to year 1479. Therefore, a possible date for the data above is March 11, 1479. As we shall see, Diego de Torres could have taken all these data from his neighbor, Abraham Zacut. In the first place, it can be deduced from the *Ḥibbur* that the entry of the Sun into Aries occurred that year on March 11 at 2;55h p.m. at Salamanca. Then, using the tables in the *Ḥibbur* for the planetary positions at that time at Salamanca, we find:

Sun	Ari	0; 0°
Mercury	Ari	15;11°
Venus	Psc	25;42°
Mars	Ari	24;24°
Lunar node	Cnc	28;10°
Moon	Sco	14;19°

Except for the Moon, the computed values agree with those given by Diego de Torres. This is not surprising for, in chapter 7 of his *Obra astrológica,* Diego de Torres refers to an "almanach" for finding the ascendant: *busca luego para esto el grado del sol por tus tablas o almanach y vete con aquel grado . . .* (f. 159v). Moreover, in chapter 1 he refers to a table for the solar declination where the entry for 30° is 11;32° and that for 60° is 20;15° (f. 154v): these are the values found in the table for the solar declination in Zacut's *Ḥibbur*, where the maximum is 23;33°. Diego de Torres does not mention Abraham Zacut anywhere in his *Obra astrológica*; on the other hand, he does cite Nicholaus Polonius who had previously held the chair of astronomy at the University of Salamanca.

5.4.2 The impact of the *Almanach Perpetuum*

Valentim Fernandes (fl. 1494–1518) is the editor of a *Reportório dos Tempos* (Lisbon, 1518; reprinted in 1521, 1528, and several times afterwards). According to Fontoura (1983, pp. 101, 412) he was of German origin, and died before May 1519. His *Reportório* was translated from a previous work in Castilian by Andrés de Li, entitled *Reportorio de los Tiempos*, first printed in 1492 at Zaragoza (Chabás and Roca 1985, p. 95; Delbrugge 1999). Fernandes's *Reportório* also contains a "Regimento da declinação do sol (. . .) A qual declinação foi tirada puntualmente del Zacuto pello honrrado Gaspar Nicolas mestre suficiente nesta arte." According to Bensaude (1912, p. 173), Fernandes's "regimento" (or rutter) was intended for navigators, and it included instructions for using some tables extracted from Zacut's *Almanach Perpetuum* by Master Gaspar Nicolas.

André Pires is only known as the author of a *Livro de Marinharia* extant in Paris, Bibliothèque Nationale de France, MS 44340 (MS Port. 40), ff. 1–37. This work is a compilation of treatises on navigation, tables, "regiments" (or rutters), and practical instructions for navigation and, according to Albuquerque (1963, p. 31), most of the material in it was written between 1500 and 1520. It also contains two sets of four-year solar tables giving the daily longitudes and declinations of the Sun. Albuquerque has shown that these tables were based on the corresponding tables in the *Almanach Perpetuum* for various years between 1494 and 1551. The text explaining the use of this tables mentions neither Zacut nor the years for which they are valid.

The Portuguese astronomer and cosmographer, Pedro Nunes (1502–1578), published *Tratado da sphera com a theorica do sol e da lua* (Lisbon, 1537) at the end of which he included some tables giving the daily solar positions for 4 years (1537–1540). The text following these tables reads (f. 176):

> O lugar do zodiaco onda esta ho sol se sabera pellas tauoas acima postas nos quatro annos pera que sam feitas (. . .) e por cada reuolução acrecentaremos hun minuto & quarenta e seys segundos ao que acharnos no tauoa: pera termos o verdadeiro lugar do sol.[3]

The rule given here, add 0;1,46° for each 4-year cycle, is taken from the *Almanach Perpetuum*, where it is presented in tabular form (Table AP 4). Moreover, it can be shown that the preceding tables for years

[3] Nunes's *Tratado da sphera* is available in a modern reprint edition: see Nunes 1940 in the bibliography. The tables and the relevant passage are in vol. 1, pp. 234–238. The relationship of this passsage in Nunes's treatise with the *Almanach Perpetuum* was pointed out by J. Bensaude (1917, p. 69).

1537–1540 are close to those derived from Zacut's tables for the daily solar positions for years 1473–1476 (Table AP 2), by comparing the corresponding entries in both tables. The time difference between these tables is 64 years (exactly 16 cycles of 4 years each); adding Zacut's entries for 1473 the corresponding correction for 16 cycles, one gets Nunes's entries for 1537 but for a constant, 0;3°. The difference in longitude between Salamanca and Lisbon is 25;46° − 22;54° = 2;52° (corresponding to about 0;11h), using data available to fifteenth-century astronomers, or 5;40° W − 9;8° W = −3;28° (corresponding to about −0;14h), using modern data. It is clear that none of these differences fully explains the systematic difference in solar motion of 0;3° (corresponding to about 1;13h) mentioned above, for which we are not able to offer a better explanation. Nunes does not cite Zacut in this passage and, as far as we can tell, he does not cite Zacut anyhere in his published works.

If the following episode described by the famous humanist and philologist, Nicolaas Beken Cleynaerts (or, Clenardus, in Latin) refers to Abraham Zacut, it indicates how widespread this astronomer's fame was in the sixteenth century. This passage has generally been taken as a story in which Abraham Zacut is mentioned, but it might be about a different Zacut. Cleynaerts was born in 1495 in Brabant, and taught Greek and Hebrew at the University of Leuven, and later at the University of Salamanca. Cleynaerts died in 1542 in Granada and, shortly before, while sojourning in North Africa, he wrote from Fez several letters to João Petit, Bishop of Santiago, Cabo Verde, which were printed, together with the rest of his correspondence in *Nic. Clenardi Epistolarum Libri Duo* (Antwerp, 1566). In a letter dated August 21, 1541 Cleynaerts tells the following story to the Bishop (p. 201; see also Roersch 1940–1941, I:190–193 and III:144):

> Audi et aliud. Medicus quo usus est Gulielmus febricitans, talis est Astrologus, praeceptore usus Zacuto, ut hodie parem non habeant Iudaei. Is facta mentione huius scientiae, quam ego plane futilem praedicabam. Et quotum, inquit, annum agis? Ut scilicet de fatis meis aliquid pronunciaretur.
>
> Listen to something else. The doctor in charge of William while he was feverish is an astrologer whose master was Zacut, and no one surpasses him nowadays among the Jews. He [the doctor], when mentioning this science, which I considered worthless, asked me: "How old are you?," evidently in order to say something about my fate.

A contemporary of Cleynaerts, Henri Baers (also known as Vekenstyl) was an instrument-maker who wrote some short astronomical treatises and published a set of tables in Leuven (1528), entitled: *Tabule perpetue longitudinum ac latitudinum planetarum noviter copulate ad meridiem universitatis Lovaniensis ac plerumque aliorum necessarium in nativitatibus requisitorum Lovanij noviter impresse*. Poulle and De Smet (1976) published a facsimile edition with a French translation and commentary. These tables are accompanied by brief explanations for their use, and the chapter for nativities (*Pro inventione gradus ascendentis nativitatis*) was taken almost word-for-word from the *Canon ultimus de animodar* in the Latin version of the *Almanach Perpetuum*. Moreover, the table for the *nimodar* (b1v-b2r) reproduces that in the *Almanach Perpetuum* (see Table AP 46), probably taken from the 1525 edition, without acknowledgment.

Once again in Salamanca, but now at the beginning of the sixteenth century, we find new evidence for the impact of the tables of Zacut. At the Biblioteca Universitaria de Salamanca there is a volume (shelfmark: 38678) containing two printed books by Alfonso de Córdoba: his annotated edition of the *Almanach Perpetuum* (1502) and the *Tabule Astronomice Elisabeth Regine* (1503), both printed in Venice by Petrus Liechtenstein. Between the two books, and bound together with them, there is a long hand-written text in Latin, as well as some tables, with numbered comments referring to the tables of the *Almanach Perpetuum*, explaining their use and occasionally adding descriptions of them (ff. 243r-266r). The comments add no new information about the computations that underly the tables, but the numerous technical details given by the author indicate that he had a real insight into their use. Although this text does not follow exactly the sequence of the printed canons, all of them are commented upon and some of the comments are even longer than the corresponding canons. Moreover, a few of the comments are heavily annotated in the margins by the same hand, and there are various references to other authors with quite a number of worked examples.

The colophon gives the date for the completion of this commentary as July 18, 1509, but no place is indicated. This means that some of the marginal notes were written subsequently, for they mention Pedro Sánchez Ciruelo's *Apotelesmata Astrologiae Christianae* (Alcalá de Henares, 1521), Johannes Stoeffler's *Elucidatio fabricae ususque astrolabii* (Oppenheim, 1513) and his *Calendarium Romanum Magnum* (Oppenheim, 1518), among other later works. The author refers several times to the edition of the *Almanach Perpetuum* "printed in Castile" (*in zacuto impreso castele*), but this is probably a confusion between Leiria, Portugal, and Salamanca. Among the printed books on astronomy prior to 1509 cited in this commentary we find: The Alfonsine Tables, Bianchini's Tables, Regiomontanus's *Kalendarium* and *Tabule Directionum*, John of Saxony's commentary on al-Qabīṣī, and Ciruelo's *Dialogue on Sacrobosco's Sphere*.

The worked examples in this commentary give priority to one date, namely, September 8, 1499, and on f. 245v the nativity of someone (the author?) born on September 8, 1473, at 2 p. m., is mentioned. Among the learned persons, well informed on astronomy, in Salamanca at the beginning of the sixteenth century, there are many candidates to be considered as the author of this commentary. We will just mention one: the author of various treatises on mathematics and logic, Pedro Sánchez Ciruelo, a mathematician born in Zaragoza *ca.* 1470, who studied at the University of Salamanca, then spent 10 years at the University of Paris (1492–1502), and the last 10 years of his life in Salamanca, where he died in 1548.

Aguirre (1980, p. 123) suggests that Zacut is referred to as "Abraham de seculo" in Zarzoso's treatise on an equatorium (Paris 1526, f. 20r) in a passage concerning the lunar mansions. But this is by no means certain, for we note that, according to Zarzoso, a lunar mansion is 12;51,26°, whereas in Zacut's *Tratado breve en las ynfluencias del cielo* (Carvalho 1927, p. 34) a lunar mansion is 12;51,28°.

The echoes of Zacut's fame among Spanish astronomers go deep into the sixteenth century. In a book by Pérez de Vargas (1563), the author mentions "çacuto en el almanach" (f. VIr) in regard to the time spent by the fetus in the mother's womb. This is indeed the subject of the last chapter in the *Almanach Perpetuum* in both the Castilian and the Latin versions.

5.5 The *Almanach Perpetuum* in the Muslim world

There are at least two versions of the *Almanach Perpetuum* in Arabic. We have already discussed the first version in chap. 5.3, for it was probably addressed to members of the Jewish community.

A second version, with both canons and tables, was composed in the Maghrib by Aḥmad b. Qāsim al-Ḥajarī al-Andalusī (b. *ca.* 1569, d. after 1641) from the Castilian version printed in 1496. According to his autobiography, al-Ḥajarī grew up in a town called al-Ḥajar al-Aḥmar ("the red stone") in Spain; this place has not been identified, but it was probably in either Andalucía or Extremadura. He lived in Seville before reaching Morocco in about 1599 where he resided in Marrakesh. He then visited a number of cities in France and Holland, and in about 1634 he left Marrakesh on a pilgrimage to Mecca. He was in Egypt in 1637 and in Tunis until 1641, after which time nothing is known about him (Koningsveld et *al.* 1997, pp. 18, 28, 49, 55).

Al-Ḥajarī's version is extant in several MSS: Milan, Ambrosiana, MS Ar. 338 (formerly C82: Griffini 1916, pp. 88–106); Vatican, MS Ar. 963, described as a copy of Milan, Ambrosiana, MS Ar. 338 (Levi della Vida 1935, pp. 101–102); and Rabat, Ḥasaniyya, MSS Ar. 1433 and 8184. Brief catalogue descriptions of two other manuscripts suggest that there are additional copies: Rabat, Ḥasaniyya, MS Ar. 1331 (Khattabi 1983, pp. 231–232, 284–287); and Milan, Ambrosiana, MS Ar. 351 L, described as an "anonymous zij for the year 1472 of the Christian era", copied in

1683 (Löfgren and Traini 1975, 1:189). Cairo, Dār al-Kutub, *Mīqāt,* MS Ar. 1081, dated ca. 1200 Hijra (1785 A.D.), has the tables and three sets of canons, one of which seems to have much in common with al-Ḥajarī's text. According to King (1986, p. 140: F31), the first set of canons is anonymous, the second in 24 chapters was translated from the Spanish by Aḥmad b. Qāsim al-Jaḥdarī al-Andalusī, and the third in 8 chapters is attributed to ʿAbd al- Raḥmān al-Fāsī. The colophon to the second set (p. 19) is virtually identical with the colophon in Rabat, MS 8184, and apparently the name al-Jaḥdarī (or: al-Ḥajdarī) is just a corruption of al-Ḥajarī. A detailed description of Milan, MS 338, is given by Griffini, and a list of the chapter headings is given by Khattabi (pp. 284–285) in his description of Rabat, MS 8184: a comparison with the printed Castilian version leaves no doubt that these two manuscripts are copies of the same translation.

The only manuscripts we have seen of the Arabic versions are: Escorial, MS Ar. 966; and Cairo, MS Ar. 1081. For the other copies we have depended on catalogue descriptions. Clearly, it would be useful for a specialist to investigate these manuscripts in order to establish their relationship to one another.

Appendix 1. Zacut's *Judgments of the Astrologer*

(Translation by B. R. Goldstein of the Hebrew text in Roth 1949, pp. 447–448.)

Judgments of the astrologer, [by] the great scholar, R. Abraham Zacut, may his memory be for a blessing.

[1] Year [5]278, 29 Sivan [June 8, 1518], the Sun will be eclipsed; it indicates great changes and that peace and agreements between kings and peoples will not be fulfilled; and sour fruits will be blighted; and it indicates hatred between peoples, brothers, and loved ones; and gout (*ḥoli ha-niqris*);[1] and locusts that will destroy the wheat in some countries; and woe for Christians, especially in Spain.

[2] Year [5]280, the night of 14 Kislev [Nov. 6/7, 1519], the Moon will be eclipsed; it indicates massacres in the east; people will fight each other; diseases for good [people]; divorcing of wives; deceits and lies, and each man will lie to his fellow; woes in the lands of Islam [*lit.* Ishmael]; and wars and woes will continue until the passing of year [5]282 [1521–1522] when the survivor will be able to say, "on this day I was (re-)born"; and Israel must repent completely and pray to God that He save them from woes and wars, for all who call upon God will be rescued. These are the pangs of the Messiah, and at that time 927 years, 6 months and 2 days will be completed according to the reckoning of the Muslims [*lit.* Ishmaelites] which are lunar years, and they are equivalent to 900 solar years.

[3] In year [5]284 [1524] there [will be] a conjunction of the planets unlike any that came before it.[2] It indicates that there will be very great woes in Christian countries [*lit.* the lands of Edom] to the west, and that the sea will rise [*lit.* enter] and some of their lands will be flooded. Happy is he who waits and reaches that year in repentence, upright in heart, and in good deeds. And in that year will be redemption and salvation for Israel, even though there will be convulsions and wars until year [5]289 [1528–1529].

[1] The expression *ʿillat al-niqris* occurs in Arabic in al-Bīrūnī's *Book of Instruction in the Art of Astrology* (Wright 1934, paragraph 431) in a passage concerning diseases associated with Saturn, and it is translated there as "gout."

[2] According to Tuckerman (1964, p. 780), there was a conjunction of Mars, Jupiter, and Saturn in Pisces in Feb. 1524.

At that time [Feb. 1524] the conjunction will be between Aquarius and Pisces,[3] as took place when [Israel] entered the land in the days of Joshua and Ezra. Then there will be two eclipses,[4] and Mars will be with Saturn and Jupiter in the same conjunction; it indicates great wars like the wars of Gog and Magog, and the Messiah son of Joseph will be killed; and, since Venus is close to them,[5] there will blossom forth salvation for Israel and the coming of the Messiah son of David, may God be blessed for the sake of His name, may He help us and support us and maintain us with His righteous right hand, and may He inscribe us for a good life with all that is written for life, to look upon the goodness of God in the land of life. Amen, may it be His will.

Copied from the commentary on the prophecies of Naḥman, may peace be upon him.

[3] This expression is not standard in an astrological text. Perhaps a longitude near 330°, the boundary between Aquarius and Pisces, is intended.

[4] There was a solar eclipse on Feb. 4, 1524, and a lunar eclipse on Feb. 19, 1524.

[5] Venus was in Pisces in Feb. 1524.

Bibliography

Aguirre, A. 1980. *El Astrónomo Cellense Francisco M. Zarzoso (1556)*. Teruel.

Albuquerque, L. de. 1963. *O Livro de Marinharia de André Pires*. Lisbon.

Albuquerque, L. de. 1986. *Almanach Perpetuum de Abraão Zacuto*. Lisbon.

Albuquerque, L. de. 1988. *Astronomical Navigation*. Lisbon.

Albuquerque, L. de. 1991. *Historia de la Navegación Portuguesa*. Madrid.

Almeida, Fortunato de. 1967. *História da Igreja em Portugal*. Nova edição preparada e dirigida por Damião Peres. Porto.

Amasuno, M. V. 1972. *Un texto médico-astrológico del siglo XV "Eclipse del Sol" del Licenciado Diego de Torres*. Salamanca.

Arranz, L. 1985. *Cristóbal Colón. Diario de a bordo*. Madrid.

Barbosa, A. 1928. "O Almanach Perpetuum de Abraham Zacuto e as Tábuas Náuticas Portuguesas," *O Instituto*, 75:541–562.

Barros, João de. *Décadas da Ásia*. Lisbon. (*Décadas I-III*: 1552–1563, *Década IV*: 1615).

Beaujouan, G. 1966. "Science livresque et art nautique au XVe siècle," *Les aspects internationaux de la découverte océanique aux XVe et XVIe siècles: Actes du Ve colloque international d'histoire maritime*, S.E.V.P.E.N (Paris), 61–85; reprinted in G. Beaujouan, *Science médiévale d'Espagne et d'alentour* (Aldershot, 1992), essay IX.

Beaujouan, G. 1967. "La science en Espagne aux XIVe et XVe siècles," *Conférence du Palais de la Découverte*, D 116 (Paris), 5–45; reprinted in G. Beaujouan, *Science médiévale d'Espagne et d'alentour* (Aldershot, 1992), essay I.

Beaujouan, G. 1969. "L'astronomie dans la péninsule ibérique à la fin du moyen âge," *Agrupamento de estudos de cartografia antiga*, XXIV:3–22; reprinted in G. Beaujouan, *Science médiévale d'Espagne et d'alentour* (Aldershot, 1992), essay X.

Benjamin, F. S., Jr., and G. J. Toomer. 1971. *Campanus of Novara and Medieval Planetary Theory*. Madison.

Bensaude, J. 1912. *L'astronomie nautique au Portugal à l'époque des découvertes*. Bern; reprinted Amsterdam, 1967.

Bensaude, J. 1917. *Histoire de la Science Nautique Portugaise: Résumé*. Geneva.

Bensaude, J. 1919. *Almanach Perpetuum Celestium Motuum (Radix 1473)*. Geneva.

Beit-Arié, M., and M. Idel. 1979 (5739 A.M.). "Treatise on Eschatology and Astrology by R. Abraham Zacut," *Kiryat Sefer*, 54:174–194 [in Hebrew].

Bīrūnī. *Tafhīm*. *See* Wright 1934.

Boffito, G., and C. Melzi d'Eril. 1908. *Almanach Dantis Aligherii sive Profhacii Judaei Montispessulani*. Florence.

Cantera Burgos, F. 1931. "El judío salmantino Abraham Zacut," *Revista de la Academia de Ciencias Exactas, Físicas y Naturales, Madrid*, 27:63–398.

Cantera Burgos, F. 1935. *Abraham Zacut*. Madrid.

Cantera Burgos, F. 1959. "Nueva serie de manuscritos hebreos de Madrid," *Sefarad*, 19:3–47.

Carvalho, J. de. 1927. "Dois inéditos de Abraham Zacuto," *Revista de Estudos Hebráicos*, 1:7–54.

Carvalho, J. de. 1947. "Dois inéditos de Abraham Zacuto," *Estudos sobre a Cultura Portuguesa do século XVI*, vol. I, pp. 95–108.

Chabás, J. 1991. "The Astronomical Tables of Jacob ben David Bonjorn," *Archive for History of Exact Sciences*, 42:279–314.

Chabás, J. 1992. *L'astronomia de Jacob ben David Bonjorn*. Barcelona.

Chabás, J. 1996a. "Astronomía andalusí en Cataluña: las Tablas de Barcelona," in J. Casulleras and J. Samsó, eds., *From Baghdad to Barcelona. Studies in the Islamic Exact Sciences in Honour of Prof. Juan Vernet*. Barcelona, pp. 477–525.

Chabás, J. 1996b. "El almanaque perpetuo de Ferrand Martines (1391)," *Archives internationales d'histoire des sciences,* 46:261–308.

Chabás, J. 1998. "Astronomy in Salamanca in the Mid-Fifteenth Century: The *Tabulae Resolutae*," *Journal for the History of Astronomy*, 29:167–175.

Chabás, J., and B. R. Goldstein. 1992. "Nicholaus de Heybech and His Table for Finding True Syzygy," *Historia Mathematica*, 19:265–289.

Chabás, J., and B. R. Goldstein. 1994. "Andalusian Astronomy: *al-Zīj al-Muqtabis* of Ibn al-Kammād," *Archive for History of Exact Sciences*, 48:1–41.

Chabás, J., and B. R. Goldstein. 1997. "Computational Astronomy: Five Centuries of Finding True Syzygy," *Journal for the History of Astronomy*, 28:93–105.

Chabás, J., and B. R. Goldstein. 1998. "Some Astronomical Tables of Abraham Zacut Preserved in Segovia," *Physis*, 35:1–10.

Chabás, J., and A. Roca. 1985. *El "Lunari" de Bernat de Granollachs*, Barcelona.

Chabás, J., A. Roca, and X. Rodríguez. 1988. "Recalculació de taules de paral.laxi de finals de l'Edat Mitjana," in L. Navarro, ed., *Trobades científiques de la Mediterrània. Història de la física*. Barcelona, pp. 237–248.

Chabás, J., and A. Roca. 1998. "Early Printing of Astronomy: The *Lunari* of Bernat de Granollachs," *Centaurus*, 40:124–134.

Chabás, J., and A. Tihon. 1993. "Verification of parallax in Ptolemy's *Handy Tables*," *Journal for the History of Astronomy*, 24:123–41.

Cohen, M. A. 1965. *Samuel Usque's Consolation for the Tribulations of Israel*. Philadelphia.

Cohn, B. 1917. Review of: "*Almanach Perpetuum coelestium motuum (radix 1473)* . . . Reproduction facsimilé, Edition 1496, . . . München 1915," *Vierteljahrsschrift der Astronomischen Gesellschaft*, 25:102–123.

Cohn, B. 1918. "Der Almanach perpetuum des Abraham Zacuto," *Schriften der Wissenschaftlichen Gessellschaft in Straßburg*, 32:1–33.

Comes, M. 1991. "Deux échos andalous à Ibn al-Bannā' de Marrākush," *Actes du VIIe Colloque Universitaire Tuniso-Espagnol sur le Patrimoine Andalous dans la Culture Arabe et Espagnole*. Tunis.

Dalen, B. van. 1993. *Ancient and Medieval Astronomical Tables: mathematical structure and parameter values*. Utrecht.

David, A. 1992. "The Spanish exiles in the Holy Land" in H. Beinart, ed., *The Sephardi Legacy*, 2 vols. Jerusalem, 2:77–108.

Delbrugge, L. 1999. *Andrés de Li. Reportorio de los tiempos*. London.

Derenbourg, H. 1941. *Les manuscrits arabes de l'Escurial*, tome II, fasc. 3: Sciences exactes et sciences occultes. Paris.

Dobrzycki, J. 1987. "The *Tabulae Resolutae*," in M. Comes, R. Puig, and J. Samsó, *De Astronomia Alphonsi Regis*. Barcelona, pp. 71–77.

Eisner, S. 1980. *The Kalendarium of Nycholas de Lynn*. London.

Flórez, C., et al. 1989. *La Ciencia del Cielo*. Salamanca.

Fontoura da Costa, A. 1983. A *marinharia dos descubrimientos*, 4th edn. Lisbon.

Freimann, Alfred. 1920. "Die Ascheriden (1267–1391)," *Jahrbuch der Jüdisch-Literarischen Gesellschaft*, 13:142–254.

Freimann, Alfred. 1924. *A. Zacut, Sefer Yuḥasin* [*The Book of Genealogies*]. Frankfurt a/M.

Freimann, Aron. 1925. "Die hebräischen Inkunabeln der Druckereien in Spanien und Portugal," in A. Ruppel, ed., *Gutenberg Festchrift zur Feier des 25 Iaehrigen Bestehens des Gutenbergmuseums in Mainz*. Mainz, pp. 203–206.

Gandz, S. 1956. *Maimonides: Sanctification of the New Moon*. Translated by S. Gandz, with supplementation by J. Obermann, and an astronomical commentary by O. Neugebauer. New Haven.

Góis, Damião de. 1690. *Crónica de El-Rei D. Manuel*. Coimbra.

Goldberg, B., and L. Rosenkranz (eds.). 1846–48. *R. Isaac Israeli: Liber Jesod Olam seu Fundamentum Mundi*. Berlin.

Goldstein, B. R. 1967. *Ibn al-Muthannā's Commentary on the Astronomical Tables of al-Khwārizmī*. New Haven.

Goldstein, B. R. 1974. *The Astronomical Tables of Levi ben Gerson*. Hamden.

Goldstein, B. R. 1979. "The Survival of Arabic Astronomy in Hebrew," *Journal for the History of Arabic Science*, 3:31–39.

Goldstein, B. R. 1980. "Solar and Lunar Velocities in the Alfonsine Tables," *Historia Mathematica*, 7:134–140.

Goldstein, B. R. 1981. "The Hebrew Astronomical Tradition: New Sources," *Isis*, 72:237–251.

Goldstein, B. R. 1987. "Descriptions of Astronomical Instruments in Hebrew," *Annals of the New York Academy of Sciences*, 500:105–141.

Goldstein, B. R. 1994. "Historical Perspectives on Copernicus's Account of Precession," *Journal for the History of Astronomy*, 24:189–197.

Goldstein, B. R. 1998. "Abraham Zacut and the Medieval Hebrew Astronomical Tradition," *Journal for the History of Astronomy*, 29:177–186.

Goldstein, B. R., and J. Chabás. 1996. "Ibn al-Kammād's Star List," *Centaurus*, 38:317–334.

Goldstein, B. R., and J. Chabás. 1999. "An Occultation of Venus Observed by Abraham Zacut in 1476," *Journal for the History of Astronomy*, 30:187–200.

Griffini, E. 1916. "Lista dei manoscritti arabi nuovo fondo della Biblioteca Ambrosiana di Milano," *Rivista degli Studi Orientali*, 7:51–130.

Jardine, L. 1996. *Worldly Goods: A New History of the Renaissance*. New York.

Kaufmann, D. 1897. "La prétendue signature d'Abraham Zacouto," *Revue des Études Juives*, 34:120–121.

Kayserling, M. 1896. "Notes sur l'histoire des Juifs au Portugal," *Revue des Études Juives*, 32:282–284.

Kennedy, E. S. 1956. "A Survey of Islamic Astronomical Tables," *Transactions of the American Philosophical Society*, vol. NS 46.

Kennedy, E. S. 1988. "Two Medieval Approaches to the Equation of Time," *Centaurus*, 31:1–8.

Kennedy, E. S. 1995/96. "Treatise V of Kāshi's *Khāqāni Zīj*: The Determination of the Ascendent," *Zeitschrift für Geschichte der Arabisch-Islamischen Wissenschaften*, 10:123–145.

Khattabi, M. A. 1983. *Catalogues of the Al-Hassania Library*, vol. 3: Manuscripts of Mathematics, Astronomy, Astrology and Geography. Rabat. [in Arabic]

King, D. A. 1986. *A Survey of the Scientific Manuscripts in the Egyptian National Library*. Winona Lake, Indiana.

King, D. A. 1987. "Some Early Islamic Tables for Determining Lunar Crescent Visibility," *Annals of the New York Academy of Sciences*, 500:185–225.

Knorr, W. R. 1997. "Sacrobosco's *Quadrans*: Date and Sources," *Journal for the History of Astronomy*, 28:187–222.

Koningsveld, P. S. van, et al. 1997. Aḥmad ibn Qāsim al-Ḥajarī, *Kitāb Nāṣir al-Dīn ʿalā 'l-Qawm al-Kāfirīn*, historical study, critical edition, and annotated translation by P. S. van Koningveld, Q. al-Samarrai, and G. A. Wiegers. Madrid.

Kunitzsch, P. 1986–1990. *Der Sternkatalog des Almagest: Die arabisch-mittelalterliche Tradition.* 2 vols. Wiesbaden.

Laguarda, R. A. 1990. *La Ciencia Española en el Descubrimiento de América.* Valladolid.

Langermann, Y. T. 1988. "The Scientific Writings of Mordekhai Finzi," *Italia*, 7:7–44.

Langermann, Y. T. 1999. "Science in the Jewish Communities of the Iberian Peninsula: An Interim Report," Essay I in Y. T. Langermann, *The Jews and the Sciences in the Middle Ages*. Aldershot.

Lay, J. 1991. *L'Abrégé de l'Almageste, attribué à Averroès, dans sa version hébraïque*. Thèse de doctorat (nouveau régime), École Pratique des Hautes Études: Section des Sciences Religieuses. Paris.

Lay, J. 1996. "L'*Abrégé de l'Almageste*: un inédit d'Averroès en version hébraïque," *Arabic sciences and philosophy*, 6:23–61.

Levi della Vida, G. 1935. *Elenco dei Manoscritti arabi islamici della Biblioteca Vaticana*, Studi e Testi, 67. Vatican City.

Levy, R. 1928. Review of: "Dois inéditos de Abraham Zacuto" by J. de Carvalho, *O Instituto*, 76:392–394.

Levy, R. 1932. "A propos de la *Magna Compositio* de Zacuto," *Revue des études juives*, 92:175–178.

Levy, R. 1936. "Zacuto's Astronomical Activity," *Jewish Quarterly Review*, 26:385–388.

Löfgren, O., and R. Traini. 1975. *Catalogue of the Arabic Manuscripts in the Bibliotheca Ambrosiana*, vol. 1. Vicenza.

Lowry, M. 1979. *The World of Aldus Manutius: Business and Scholarship in Renaissance Venice*. Ithaca.

Lucena e Vale, Alexandre de. 1934. *O Bispo de Viseu D. Diogo Ortiz de Vilhegas. O cosmógrafo de D. João II*. Gaia.

Luzzatto, A. 1972. *Hebrew Ambrosiana: Catalogue of Undescribed Hebrew Manuscripts in the Ambrosiana Library*. Milan.

Maddison, F. 1992. "A Consequence of Discovery: Astronomical Navigation in

Fifteenth-Century Portugal" in T. F. Earle and S. Parkinson, eds., *Studies in the Portuguese Discoveries I*. Warminster, pp. 71–110.

Mahler, E. 1916. *Handbuch der jüdischen Chronologie*. Leipzig.

Maimonides. *See* Gandz 1956.

Margoliouth, G. 1915. *Catalogue of the Hebrew and Samaritan Manuscripts in the British Museum, Part III*. London.

Mestres, A. 1996. "Maghribī Astronomy in the 13th Century: a Description of Manuscript Hyderabad Andra Pradesh State Library 298," in J. Casulleras and J. Samsó, eds., *From Baghdad to Barcelona. Studies in the Islamic Exact Sciences in Honour of Prof. Juan Vernet*. Barcelona, pp. 383–443.

Millás, J. M. 1942. *Las traducciones orientales en los manuscritos de la Biblioteca Catedral de Toledo*. Madrid.

Millás, J. M. 1943–1950. *Estudios sobre Azarquiel*. Madrid-Granada.

Millás, J. M. 1946. "Un tratado de almanaque probablemente de R. Abraham ibn ʿEzra," *Studies and Essays in the History of Science and Learning Offered in Homage to George Sarton,* ed. by M. F. Ashley Montagu. New York, pp. 421–432.

Millás, J. M. 1947. *El libro de los fundamentos de las Tablas astronómicas de R. Abraham ibn ʿEzra*. Madrid-Barcelona.

Millás, J. M. 1962. *Las Tablas Astronómicas del Rey Don Pedro el Ceremonioso*. Madrid-Barcelona.

Nallino, C. A. 1899–1907. *Al-Battānī sive Albatenii Opus Astronomicum*, 3 vols. Milan.

Neubauer, A. 1886. *Catalogue of the Hebrew Manuscripts in the Bodleian Library*. Oxford.

Neugebauer, O. 1962. *The Astronomical Tables of al-Khwārizmī*. Copenhagen.

Neugebauer, O. 1975. *A History of Ancient Mathematical Astronomy*. Berlin-New York.

North, J. D. 1977. "The Alfonsine Tables in England," in Y. Maeyama and W. G. Salzer, eds., *Prismata: Festschrift für Willy Hartner*. Wiesbaden, pp. 269–301.

North, J. D. 1986. *Horoscopes and History*. London.

Nunes, P. 1537. *Tratado da sphera com a theorica do sol e da lua*. Lisbon; reprinted in P. Nunes *Obras* (Lisbon, 1940), 3 vols.

Oppolzer, T. von. 1887. *Canon der Finsternisse*. Vienna.

Pérez de Vargas, B. 1563. *Segunda Parte de la Fábrica del Universo llamada Repertorio Perpetuo*. Toledo.

Peurbach, G. 1514. *Tabulae eclypsium*. Vienna.

Porres, B. and J. Chabás. 1998. "Los cánones de las *Tabulae Resolutae* para Salamanca: origen y transmisión." *Cronos*, 1:51–83.

Poulle, E. 1969. "Les conditions de la navigation astronomique au XVe siècle," *Revista da Universidade de Coimbra*, 24:3–20.

Poulle, E. 1973. "John of Lignères" in *Dictionary of Scientific Biography*, 7:122–128.

Poulle, E. 1984. *Les Tables Alphonsines avec les canons de Jean de Saxe*. Paris.

Poulle, E. and A. De Smet. 1976. *Les Tables astronomiques de Louvain de 1528 par Henri Baers ou Vekenstyl*. Brussels.

Ptolemy. *Almagest*. *See* Toomer 1984.

Ptolemy. *Handy Tables*. *See* Stahlman 1959.

Ratdolt, E. (ed.) 1483. *Tabulare astronomice illustrissimi Alfontij regis castelle*. Venice.

Rico y Sinobas, M. 1863–1867. *Libros del Saber de Astronomía del Rey D. Alfonso X de Castilla*, 5 vols. Madrid.

Robbins, F. E. (ed. and trans.). 1940. *Ptolemy: Tetrabiblos*. London and Cambridge, Mass.

Roersch, A. 1940–1941. *Correspondance de Nicolas Clénard*. 3 vols. Brussels.

Rome, A. 1931. *Commentaires de Pappus et de Théon d'Alexandrie sur l'Almageste*, t. I: *Pappus d'Alexandrie. Commentaire sur les livres 5 et 6 de l'Almageste*. Rome.

Rose, P. L. 1975. *The Italian Renaissance of Mathematics*. Geneva.

Rosińska, G. 1984. *Scientific Writings and Astronomical Tables in Cracow: A Census of Manuscript Sources (XIVth-XVIth Centuries)*. Wrocław.

Roth, C. 1936–1937. "A Note on the Astronomers of the Vecinho Family," *Jewish Quarterly Review*, 27:233–236.

Roth, C. 1937–1938. "A Further Note on the Astronomers of the Vecinho Family," *Jewish Quarterly Review*, 28:157.

Roth, C. 1949. "The Last Years of Abraham Zacut," *Sefarad*, 9:445–454.

Roth, C. 1954. "Who Printed Zacuto's Tables?," *Sefarad*, 14:122–125.

Saby, M.-M. 1987. *Les canons de Jean de Lignères sur les tables astronomiques de 1321*. Unpublished thesis: Ecole Nationale des Chartes, Paris. A summary appeared as: "Les canons de Jean de Lignères sur les tables astronomiques de 1321," *Ecole Nationale des Chartes: Positions des thèses*, 1987, pp. 183–190.

Samsó, J. 1994. "Trepidation in al-Andalus in the 11th Century," Essay VIII in *Islamic Astronomy and Medieval Spain*. Aldershot.

Samsó, J. 1997. "Andalusian Astronomy in 14th Century Fez: *al-Zīj al-Muwāfiq* of Ibn ʿAzzūz al-Qusanṭīnī," *Zeitschrift für Geschichte der Arabisch-Islamischen Wissenschaften*, 11:73–110.

Samsó, J., and E. Millás. 1998. "The Computation of Planetary Longitudes in the *Zīj* of Ibn al-Bannā' ", *Arabic Sciences and Philosophy*, 8:259–286.

Sassoon, D. S. 1932. *Ohel Dawid: Descriptive Catalogue of the Hebrew and Samaritan Manuscripts in the Sassoon Library, London*. London.

Schwarz, A. Z., D. S. Loewinger, and E. Roth. 1973. *Die hebräischen Handschriften in Österreich* (Texts and Studies, American Academy for Jewish Research, vol. 4). New York.

Seligsohn, M. 1904. "Galina, Moses ben Elijah," in *The Jewish Encyclopedia*, 5:554–555.

Sezgin, F. 1978. *Geschichte des arabischen Schrifttums*, Band vi: Astronomie. Leiden.

Shochat, A. 1948–1949. "Rabbi Abraham Zacuto in the Talmudical Academy of Rabbi Isaac Shulal in Jerusalem," *Zion*, 13–14:43–46 (in Hebrew; English summary, p. iii).

Solon, P. 1970. "The *Six Wings* of Immanuel Bonfils and Michael Chrysokokkes," *Centaurus* 15:1–20.

Stahlman, W. D. 1959. *The Astronomical Tables of Codex Vaticanus Graecus 1291*. Brown University, Ph. D. dissertation. University Microfilms, No. 62–5761.

Steinschneider, M. 1964. *Mathematik bei den Juden*, 2nd edn. Hildesheim.

Stillwell, M. B. 1970. *The Awakening Interest in Science during the First Century of Printing: 1450–1550*. New York.

Suter, H. 1914. *Die astronomischen Tafeln des Muḥammad ibn Mūsā al-Khwārizmī.* Copenhagen.

Swerdlow, N. M. 1973. "The Derivation and First Draft of Copernicus's Planetary Theory: A Translation of the *Commentariolus* with Commentary," *Proceedings of the American Philosophical Society*, 117:423–512.

Swerdlow, N. M. 1977. "Summary of the Derivation of the Parameters in the *Commentariolus* from the Alfonsine Tables," *Centaurus,* 21:201–213.

Toomer, G. J. 1968. "A Survey of the Toledan Tables," *Osiris*, 15:5–174.

Toomer, G. J. 1973. "Prophatius Judaeus and the Toledan Tables," *Isis*, 64:351–355.

Toomer, G. J. 1984. *Ptolemy's Almagest*. New York.

Tuckerman, B. 1964. *Planetary, Lunar, and Solar Positions A.D. 2 to A.D. 1649 at Five-day and Ten-day Intervals*. Philadelphia.

Vernet, J. 1949. "Un tractat d'obstetricia astrològica," *Boletín de la Real Academia de Buenas Letras de Barcelona*, 22:69–96; reprinted in Vernet 1979, pp. 273–300.

Vernet, J. 1950. "Una versión árabe resumida del *Almanach Perpetuum* de Zacuto," *Sefarad*, 10:115–133; reprinted in Vernet 1979, pp. 333–351.

Vernet, J. 1956. "*Las Tabulae Probatae*," *Homenaje a Millás Vallicrosa*, 2 vols. Barcelona, 2:501–522; reprinted in Vernet 1979, pp. 191–212.

Vernet, J. 1979. *Estudios sobre Historia de la Ciencia Medieval*. Barcelona-Bellaterra.

Viterbo, S. 1898–1900. *Trabalhos náuticos dos Portugueses. Séculos XVI e XVII*, 2 vols. Lisbon; reprinted Lisbon 1988.

Wright, R. Ramsay (ed. and trans.). 1934. *The Book of Instruction in the Elements of the Art of Astrology* by Abu'l-Rayḥān Muḥammad ibn Aḥmad al-Bīrūnī. London.

Zacut, A. *Sefer Yuḥasin*. *See* Alfred Freimann 1924.

Zinner, E. 1990. *Regiomontanus, his life and work*. Translated by E. Brown. Amsterdam.

NOTATION

α	true anomaly
$\bar{\alpha}$	mean anomaly
β	celestial latitude
δ	declination
Δ	increment
ε	obliquity of ecliptic
η	elongation
κ	(argument of) center
λ	celestial longitude
ϕ	geographical latitude
ω	argument of latitude

d	digit, or day
D	magnitude of an eclipse
h	hour
i	inclination of the lunar orbit
L	geographical longitude
M	length of daylight
p_β	parallax in latitude
r	radius of the Moon
s	radius of the Sun, or zodiacal sign
t	time
z	radius of the Earth's shadow
v	velocity

INDICES

Manuscripts Cited

Bruges
Stadsbibliotheek
MS 522: 103

Cairo
Dār al-Kutub
Mīqāt, MS Ar. 1081: 171

Cracow
Jagiellonian Library
MS 553: 138
MS 573: 20
MS 609: 28, 152
MS 610: 28, 138
MS 613: 28
MS 1864: 20
MS 1865: 20, 28

Cusa
Bibliothek im St. Nikolaus Hospital
MS 211: 28

Erfurt
Bibliotheca Amploniana
MS Q381: 17

Escorial
Biblioteca del Monasterio
MS Ar. 939: 152
MS Ar. 966: 163, 171

Jerusalem
Benayahu MS Heb. 136: 163

Leeuwarden
MS Heb. 5: 51

Lisbon
Arquivo Nacional da Torre do Tombo
Corpo Chronologico, parte 1ª, maço 2, doc. 18: 9, 12

London
  British Library
    MS Vesp. F11: 17
    MS Ar. 977: 17
    MS Heb. Add. 27,106: 51
    MS Or. 10725: 50
  Jews College
    MS Heb. 135: 152
  Sassoon
    MS Heb. 799: 6
    MS Heb. 823: 70

Lyon
  Bibliothèque Municipale
    MS Heb. 14: 50, 53–56, 58–62, 64–66, 69–75, 78, 80–89, 101, 108, 112, 121, 123, 125–129, 141, 145, 147, 155

Madrid
  Academia de la Historia
    MS Heb. 14: 53–56, 58, 62, 65, 68–69, 73–75, 82–84, 101
  Biblioteca Nacional
    MS 3306: 18
    MS 3385: 21, 23–47, 53–60, 62–63, 68–69, 73–75, 107, 109, 117, 122, 131–132, 134, 136, 147, 150, 165
    MS 10023: 152

Milan
  Biblioteca Ambrosiana
    MS Ar. 338: 170–171
    MS Ar. 351 L: 170
    MS Heb. X-193 Sup.: 22

Munich
  Staatsbibliothek
    MS Heb. 109: 53–56, 58, 62–66, 68–70, 72–75, 77–78, 84–85, 87, 101, 141, 155
    MS Heb. 126: 22, 85
    MS Heb. 343: 22, 108, 138, 140, 152
    MS Heb. 384: 60
    MS Heb. 386: 108

New York
  Jewish Theological Seminary of America

- MS Heb. 296 (Benaim MS 44): 53, 163
- MS Hcb. 2554: 163
- MS Heb. 2597: 109
- MS Heb. 2602: 163
- MS Heb. 5782: 164
- MS Heb. 8112: 163

Oxford
- Bodleian Library
  - MS Can. Misc. 27: 20, 24–26, 28, 35–36, 140
  - MS Lyell Heb. 96: 22
  - MS Opp. Add. 8° 42: 6, 11, 34, 50–53, 81, 109, 111–112, 121–123, 126–128, 131, 141, 145, 163
  - MS Poc. 368 (Nb. 2044): 51, 63, 118
  - MS Rawlinson D 1227: 138
  - MS Regio 14: 22

Paris
- Alliance Israélite
  - MS Heb. VIII.E.60: 6
- Bibliothèque Nationale de France
  - MS Heb. 724: 142
  - MS Heb. 725: 142
  - MS Heb. 1031: 51
  - MS Heb. 1085: 51, 63, 118
  - MS 44340 (MS Port. 40): 167

Rabat
- Ḥasaniyya
  - MS Ar. 1331: 170
  - MS Ar. 1433: 170
  - MS Ar. 8184: 170–171

St. Petersburg
- Academy of Sciences
  - MS Heb. C-076: 22, 51

Salamanca
- Biblioteca Universitaria
  - MS 2-163: 6–7, 53, 84, 109, 111–112, 121–123, 126–128, 141, 145
  - MS 2662: 103

Segovia
Biblioteca de la Catedral
MS 110: 53–56, 60, 65–68, 70, 73–75, 87–89, 101, 107, 132, 144, 164

Seville
Biblioteca Colombina
MS 5-2-21: 8
MS 5-2-32: 96

Toledo
Biblioteca de la Catedral
MS 98-27: 47

Vatican
Biblioteca Apostolica
MS Heb. 384: 50, 84, 111, 122, 124, 145
MS Ar. 963: 170

Vienna
Nationalbibliothek
MS 2440: 28, 138, 140
MS 5151: 138, 140

Warsaw
ZIH
MS Heb. 245: 51, 53–56, 61–62, 71, 73, 101, 108

Parameters

1;46: 106–107, 167
1;53: 52
2;10: 40, 52
2;44: 23, 103
0;3,10,38,7,14: 30
0;3,10,38,7,14,49,10: 117
4;29: 32, 62, 122, 124, 131
4;30: 32, 123, 128, 131
4;56: 52
5;0: 32, 123, 130–131
6;38: 147, 150

12;11,26,41: 25
12;51,28: 170
13;3,53: 87
13;3,53,57,30,21: 26
13;3,53,57,30,21,4,13: 116
13;10,35: 32, 87
13;10,35,1: 25
0;13,55,30: 126
0;13,56: 128
0;14,27,45: 126

21;46: 83
23;30: 155
23;33: 66, 101, 104, 106, 122, 166
23;35: 131
25;46: 36, 83, 103, 115, 155
28;30: 36, 103, 155
29d 12; 44h: 25
29d 12;44,3,3h: 26
29d 12;44,3,20h: 26
29d 12h 793p: 80
29;31,50,8,20: 26, 80, 84

32;30: 115, 155
32;56: 32
33;0: 65, 155
0;36,59: 144
0;36,59,27,23: 144
39;50: 36
39;54: 155

41;19: 31, 33, 36, 41, 63, 101, 155
42;30: 155

0;53,20: 122
0;56,12,30: 128
0;57,30: 122
0;59,8,19,37,19: 28
0;59,8,20: 25

66;27: 66, 69
67;30: 65

91d 7;30h: 84

365d 5;49,15,59,34,3h: 106
365d 5;49,16h: 110
365;14,33,9,58: 106

11,325: 110–112

Names and Subjects

Aboab, Isaac: 6
Abraham Ibn Ezra: 17, 49, 51, 65
Ac (siglum defined): 53
Afonso V, King of Portugal: 13–14, 157
Aguirre, A.: 170, 175
Ailly, Pierre d': 158
Albuquerque, L. de: 9n, 34, 95, 102, 113n, 158n, 167, 175
Aleppo: 163
Alfonso de Córdoba: 106–107, 150, 162–163, 169
Alfonso X, King of Castile: 2, 20, 52, 130
Almanac of Tortosa. *See* Almanac of 1307
Almanac: 2, 16–18, 103, 117; of 1307: 18, 53n, 103. *See also* Azarquiel, and Jacob ben Makhir
Almeida, F. de: 13, 175
Alvarez, P. Francisco: 14
Amasuno, M. V.: 165, 175
Andalucía: 170
Animodar: 86, 98, 150, 152–153, 163–164, 169–170
Antoninus: 150
AP (siglum defined): 53
AP and HG, corresponding tables in: 54–55
Apogee. *See* Sun, Saturn, Jupiter, Mars, Venus, and Mercury
Arranz, L.: 158n, 175
Ascendant: 63, 100–102
Ascensions, right and oblique: 63–64, 131
Asher ben Yeḥiel: 49–50
Aspects, planetary: 61, 87, 129
Associated planets: 71, 147–150
Astrology: 15, 47, 63, 70, 72, 86–89, 100–102, 147, 165–166, 169–170, 173–174
Averroes: 52, 141–142, 178
Avignon: 22
Azarquiel: 17, 56, 84, 103–104, 116, 117n, 120, 130–131

B (siglum defined): 53
Baers, Henri: 168
Baghdad: 163
Bannā', Ibn al-: 17
Barbosa, A.: 106, 175
Barcelona. *See* Tables of Barcelona
Barros, J. de: 14, 157–158, 175
Base-30 tables. *See* Trigesimal notation
Batecomb (Oxford Tables): 21–22, 138, 140
Battānī, al-: 52, 56, 73, 108, 117, 120, 130–131, 141
Beaujouan, G.: 13, 48, 175
Beit-Arié, M.: 15, 175
Benjamin, F. S., Jr.: 16
Bensaude, J.: 14, 95n, 106n, 157–158, 167n, 175
Bīrūnī, al-: 87–88, 150, 173n, 175, 181
Bianchini, Giovanni: 22, 169

Boffito, G.: 17, 103, 175
Bologna: 47
Bonfils. *See* Immanuel ben Jacob Bonfils
Bonjorn. *See* Jacob ben David Bonjorn
Bradwardine, Thomas of: 47
Brahe, Tycho: 4
Burgos: 50

Calendar, Christian: 58, 81, 155–156; Hijra: 81; Jewish: 58, 76–81, 84, 98, 159–160; Persian: 81
Canons, Latin and Castilian: 95–98
Cantera Burgos, Francisco: 1, 7–8, 9n, 34, 50–51, 63, 76, 81, 84, 95–96, 101, 104, 121, 128–129, 155, 164–165, 175
Carthage: *See* Tunis
Carvalho, J. de: 8n, 150, 170, 176, 178
Castronuño: 36
Ceuta: 14
Ciruelo, Pedro Sánchez: 169–170
Cistercian numerals: 56
Clenardus. *See* Cleynaerts, Nicolaas Beken
Cleynaerts, Nicolaas Beken: 168
Cohen, M. A.: 9n, 176
Cohn, B.: 65, 90, 155, 176
Colophon (*Almanach Perpetuum*): 9–10, 156–157
Columbus, Bartholomew: 158
Columbus, Christopher: 14, 158, 162
Columbus, Ferdinand: 158
Comes, M.: 147n, 176
Conjunction: 24–25, 59, 61, 77–79, 129. *See also Molad*, and Syzygy
Copernicus: 4, 177, 181
Córdoba, Alfonso de. *See* Alfonso de Córdoba
Correia, G.: 9
Cracow: 20, 24, 28

Dalen, B. van: 108n, 176
Daniel ben Peraḥia: 164
David, A.: 15, 176
Daylight: 154
De Smet, A.: 169, 180
Declination: 56, 68–70, 103–104, 106, 164, 166
Dedication: 90–95
Deficit: 25, 27
Delbrugge, L.: 167, 176
Derenbourg, H.: 163, 176
Digits, area: 32, 63, 125, 127; digits of eclipse (magnitude): 44, 63, 67, 153
Dignity (astrological): 72, 87
Disciples. *See* Ricius, and Vizinus
Dobrzycki, J.: 21, 177
Dominical letters. *See* Sunday letters
Double argument tables: 2, 138, 145

Eclipses, equation of: 62, 118–122; lunar: 33, 45–46, 62–63, 65–66, 127–128, 153–154, 173–174; solar: 8–9, 11, 32, 44, 62, 65, 67, 125–127, 153–154, 163, 165, 173–174; solar eclipse of March 16, 1485: 8–9, 11, 165; solar eclipse of July 29, 1478: 11
Eisner, S.: 107, 177
Elongation between the Sun and the Moon: 37–38; between the Sun and the lunar node: 62
Engel, E.: 11
Enrique IV, King of Castile: 13
Epochs. Incarnation: 40, 81, 132, 136; 137 A.D. (Antoninus 1): 150; Jan. 25, 581: 147n; Hijra: 147; September 1, 1088: 103; May 31, 1252: 22, 81; March 1, 1307: 103; March 1, 1307: 103; 1361: 110; 1401 (December 31, 1400): 82–83, 118; January 1, 1461: 24, 26, 37, 40, 47–48; March 1, 1461: 30, 53, 60, 62; December 31, 1472: 23; January 1, 1473: 55, 58–59, 156; March 1, 1473: 56, 59, 74, 101, 131, 134, 136; January 1475: 68; March 1475: 69, 73–74; January 1476: 68, 73–75, 134; 1501: 163; 1513: 163
Equation of time: 29, 48, 56, 108
Equatorium: 170
Era Caesar: 50
Euclid: 47
Exaltation, planetary: 72, 87
Extremadura: 8–9, 170
Ezra, Abraham Ibn. *See* Abraham Ibn Ezra

Faces, lords of: 87–88
Farghānī, al-: 52, 141
Fernandes, Valentim: 157–158, 167
Ferrara: 159
Fez: 168
Fixed stars. *See* Spica, and Star list
Flórez, C.: 96n, 150, 176
Fontiveros, Fernando de: 165
Fontoura da Costa, A.: 96n, 156, 158n, 167, 177
Fraenkel, M.: 11–12
Freimann, Alfred: 6n, 9n, 50, 122n, 177
Freimann, Aron: 95n, 177

Galiano, Moses: 163
Gama, Vasco da: 9n, 14
Gandz, S.: 77, 80–81, 84, 177
Gascon, Abraham: 163
Gata: 8
Gemma Frisius: 107
Genealogies, Book of: 6, 9n, 11, 15, 50, 177, 181
Genoa. *See* John of Genoa
Geographical coordinates: 36, 64, 155
Gersonides. *See* Levi ben Gerson
Gmunden. *See* John of Gmunden
Goal-year texts: 16–17
Góis, Damião de: 13, 177
Goldberg, B.: 58, 177
Gout: 173
Granada: 168
Griffini, E.: 171, 177
Grzymała, Andreas: 20
Guinea: 158

Ḥajarī, al-: 170–171
Ḥayyim of Briviesca: 49–51
Hā'im, Ibn al-: 84
Herald, D.: 141n
Heybech. *See* Nicholaus de Heybech
HG (siglum defined): 53
HG and AP, corresponding tables in: 54–55
Hipparchus: 52
Hispalensis: 162
Houses, astrological: 63, 100–102
Huber, P.: 141n
Humeniz: 103

Idel, M.: 15, 175
Immanuel ben Jacob Bonfils: 2, 49, 108–109, 181
Instruments, astronomical: 2, 9, 11
Isaac al-Ḥadib: 49
Isaac ben Sid: 49
Isaac Israeli: 49, 58, 177
Isabel, Queen of Castile: 13
Isḥāq, Ibn: 84
Israel, salvation of: 15, 173–174
Iunta, L. A.: 162

Jacob Anaṭoli: 142n
Jacob ben David Bonjorn (Poel): 2, 18, 20, 24, 44–45, 49, 52, 62, 65, 110–115, 122, 125–130
Jacob ben Makhir (Profatius): 17, 49, 103, 165
Jacob ben Tibbon. *See* Jacob ben Makhir
Jacob Mizraḥi: 163
Jardine, L.: 3n, 177
Jerusalem: 4, 15, 155, 163
João II, King of Portugal: 9, 14, 157
John of Genoa: 28
John of Gmunden: 138
John of Lignères: 18, 20, 52, 141–143, 179–180
John of Murs: 18
John of Saxony: 18–19, 169
Josepe, Master: 14, 157–159
Joy, houses of: 72, 87–89
Juan Gil: 152
Judah ben Asher I (d. 1329): 49, 122n
Judah ben Asher II (d. 1391): 2, 49–50, 60, 62–63, 65, 122–124, 128, 145
Judah ben Verga: 2, 30, 49, 51, 63, 118
Judgments of the astrologer: 15, 173–174
Junta dos Mathematicos: 14,157–158
Jupiter, anomaly of: 73, 136; apogee of: 73, 82–83; center of: 73, 135; latitude of: 73, 138, 140; longitude of: 69, 133

Kammād, Ibn al-: 20, 32, 52, 84, 104, 131, 147, 150, 152–153
Kaufmann, D.: 11, 177
Kayserling, M.: 11, 177
Kennedy, E. S.: 108n, 131n, 177–178
Khattabi, M. A.: 171, 178
Khwārizmī, al-: 32, 45, 131
King, D. A.: 171, 178
Knorr, W.: 103n, 178
Koningsveld, P. S. van: 170, 178
Kunitzsch, P.: 147, 178

L (siglum defined): 53
Ladino: 164
Laguarda, R. A.: 36, 155, 178
Langermann, Y. T.: 11, 22, 50–51, 178
Lay, J.: 142n, 178
Leuven: 153, 168
Levi ben Gerson: 2, 32, 44, 49–51, 108, 117, 122, 125, 131, 142n, 161, 177
Levi della Vida, G.: 170, 178
Lexicography: 6
Li, Andrés de: 167
Libro de las tablas alfonsíes: 18, 21
Liechtenstein, P.: 162–163, 169
Lignères. *See* John of Lignères
Lisbon: 48, 51, 168
Löfgren, O.: 171, 178
Lowry, M.: 3n, 178
Lucena e Vale, A de: 13–14, 178
Luzzatto, A.: 22, 179
Lyon: 22

Ma (siglum defined): 53
Ma^{c}shar, Abū: 8
Maddison, F.: 14, 158n, 179
Maghrib: 17
Mahler, E.: 58–59, 81, 84, 179
Maimonides: 49, 80–81, 84, 177
Makhir. *See* Jacob ben Makhir
Manuel I, King of Portugal: 9, 14
Margoliouth, G.: 51, 179
Marrakesh: 17, 170
Mars, anomaly of: 73, 136; apogee of: 73, 82–83; center of: 73, 135; correction of: 74, 85, 143; latitude of: 73, 138, 140–141; longitude of: 73, 133
Martim de Boemia (Martin Behaim): 14, 157–158
Martines, Ferrand: 18
Mecca: 170
Melzi d'Eril, C.: 17, 103, 175
Menelaus: 52
Mercury, anomaly of: 75, 83, 136; apogee of: 75, 82–83; center of: 75, 136; correction of: 75, 144; latitude of: 75, 138, 141; longitude of: 74, 134; unequal motion of: 50, 75, 145–146
Messiah, pangs of: 173. *See also* Israel, salvation of
Mestres, A.: 84, 179
Millás, J. M.: 17, 47, 84, 103–104, 117n, 120n, 131n, 179
Molad: 80. *See also* Conjunction; and Calendar, Jewish
Moon, anomaly of: 26, 38, 60, 83, 115–116; anomaly, true daily motion in: 59–61; correction of: 28; daily positions of: 60, 110–114; equation of center of: 116; latitude of: 44, 60, 86, 123, 127, 130–131. *See also* Conjunction, Eclipses, Elongation, Node, and Parallax
Moveable feasts: 58–59, 156
Moysés, master: 14
Mu (siglum defined): 53
Mumtaḥan zij: 104
Murs. *See* John of Murs

Nallino, C. A.: 117, 120n, 179. *See also* Battānī, al-
Nativities. *See* Astrology
Navigation: 2–3, 9, 11, 14, 157–158, 167
Nebrija, Antonio de: 47
Neubauer, A.: 51, 179
Neugebauer, O.: 16, 32n, 77, 84, 108n, 131n, 140–141, 143n, 177, 179
Nicholaus de Heybech: 24, 28–29, 39
Node, lunar: 39, 60, 82–83, 117; solar elongation from: 29–30, 118–119
Nonagesimal: 31, 40–41
North, J. D.: 18, 21, 101n, 138, 179
Nunes, Pedro: 107, 167–168, 179
Nycholas de Lynn: 107

Obermann, J.: 77, 177
Observations, astronomical. *See* Eclipses, solar; Instruments, astronomical; Spica; and Venus
Oppolzer, T.: 67, 153, 179
Ortas, Samuel d': 3–4, 95, 156
Ortiz. Diego Ortiz de Calçadilla: 13–15, 47, 157, 178
Ortiz. Diogo Ortiz de Vilhegas. *See* Ortiz. Diego Ortiz de Calçadilla
Oxford Tables. *See* Batecomb

Parallax: 31, 33–34, 40–44, 62, 118–125, 163
Paris: 18, 20, 22–23, 170
Pérez de Vargas, B.: 170, 179
Perpignan: 115, 122, 155
Peurbach, G.: 32, 179
Piccolomini, Aeneas Silvius: 158
Pires, André: 167
Polonio. *See* Polonius
Polonius, Nicholaus: 2, 20–21, 26, 28, 34, 40, 47–48, 165
Porres, B.: 20, 24, 132, 179
Poulle, E.: 103, 111, 117, 169, 179–180
Poznan: 20
Precession: 98
Pregnancy, duration of: 86. *See also* Animodar
Printing, introduction of: 3
Profatius. *See* Jacob ben Makhir
Ptolemy: 16, 52, 130; *Almagest*: 16, 26, 32, 45, 128, 141, 147, 150, 180–181; *Centiloquium*: 150; *Handy Tables*: 31–33, 40, 73, 108, 117, 120, 127–128, 180–181; *Cosmographia*: 65; *Tetrabiblos*: 72, 87, 150

Qabīṣī, al-: 169

R (siglum defined): 90
Ratdolt, E.: 19, 103, 111, 155, 180
Regimento de Évora: 106
Regimento de Munich: 106
Regiomontanus: 4, 90, 93, 98, 155–156, 169
Ricius, A.: 7n, 161
Rico y Sinobas, M.: 18, 103, 180
Rijāl, ^{c}Alī ibn Abī: 52
Robbins, F. E.: 72, 87, 180
Robertus Anglicus: 103
Roca, T.: 1n, 122, 176
Rodríguez, X.: 122, 176

Rodrigo, Master: 14, 157
Roersch, A.: 168, 180
Rome, A.: 120, 180
Rose, P. L.: 4n, 180
Rosenkranz, L.: 58, 177
Rosińska, G.: 138, 180
Roth, C.: 15, 95n, 159, 173, 180
Rough rule: 34
Rutter: 104, 167

Saby, M.-M.: 180
Saint-Cloud, William of: 18
Salaya, Juan de: 1, 7–8, 47, 53
Samsó, J.: 17, 84, 110n, 147n, 180
Samuel, son of Abraham Zacut: 15
Samuel, father of Abraham Zacut: 6
Santritter, J. L.: 161
Sasson, Joseph: 159–160
Sassoon, D. S.: 6, 180
Saturn, anomaly of: 68, 136; apogee of: 68, 82–83; center of: 68, 135; latitude of: 69, 138–140; longitude of: 68, 83, 132–133
Saxony. *See* John of Saxony
Schwarz, A. Z.: 53, 180
Se (siglum defined): 53
Seligsohn, M.: 163, 180
Seville: 170
Sexagesimal multiplication table: 36, 75, 145
Shochat, A.: 6, 180
Signature of Abraham Zacut: 11–12
Simon ben Jonah Mizraḥi: 163
Sine table: 63
Solomon ben Davin de Rodez: 22
Solon, P.: 109, 181
Spanish Era. *See* Era Caesar
Spica, occultation of: 11, 98, 161
Stahlman, W. D.: 31n, 117, 120, 127–128, 181
Star list: 3, 68, 71, 85, 145, 147–150
Stations: 75
Steinschneider, M.: 6n, 51, 181
Stillwell, M. B.: 3n, 181
Stoeffler, Johann: 107, 169
Ṣūfī, al-: 52, 147
Sun, anomaly of: 26, 37; apogee of: 28, 74, 82–83, 103; center of: 74; correction of: 56, 106–108, 167; daily positions of: 55–56, 101–105, 168; entry into zodiacal signs: 45–46, 48–49, 56, 109–110, 166; equation of: 29, 40, 51; longitude, mean motion in: 39. *See also* Conjunction, Declination, Eclipses, Elongation, Node, and Sunset
Sunday letters: 156
Sunset, time of: 31, 41
Suter, H.: 32n, 131n, 181
Swerdlow, N.: 106n, 181
Syzygy: 25, 28, 39, 61, 113–115, 128–130. *See also* Conjunction and *Molad*

Tables in Castilian: 37–47
Tables of Barcelona: 18, 32, 106, 131
Tabulae Resolutae: 20–21, 24, 26, 28, 35, 40–41, 47–49
Tabule Verificate: 23–41, 44, 47, 60, 62–63, 109, 116–118, 122, 127
Tangier: 14
TC (siglum defined): 37
Tequfot: 56, 84–85
Term (astrological): 72, 87n
Tihon, A.: 31n, 176
Toledo: 22, 36, 81, 155, 162; longitude (or time) difference from Salamanca: 103, 111, 132, 135–136; Toledan Tables: 18, 56, 73, 108, 117, 120, 131, 140
Toomer, G. J.: 16, 18, 26, 73, 117, 120n, 131n, 140, 147, 175, 181
Torres, Diego de: 21, 36, 165–166
Tortosa, Almanac of. *See* Almanac of 1307
Traini, R.: 171, 178
Tratado breve en las ynfluencias del cielo: 8
Trigesimal notation: 53, 81–83, 98, 112
Triplicity: 72, 87–88
Tuckerman, B.: 173n, 181
Tunis: 6n, 7, 15, 161
TV (siglum defined): 23

Vekenstyl. *See* Baers, Henri
Venus, anomaly of: 74, 83, 136; apogee of: 74, 82–83; center of: 74, 135; correction of: 74, 144, 165; latitude of: 74, 138, 140–143; longitude of: 74, 83, 133–134; occulation of: 11, 124, 141–143, 177
Verga. *See* Judah ben Verga
Vernet, J.: 104, 153, 163, 181
Vienna: 90, 95
Villalpando: 36
Viseu: 14
Visibility: 72, 75
Vital, Ḥayyim: 163
Vitéz, János: 90
Viterbo, S.: 9n, 181
Vivero, Gonzalo de: 7–8, 95
Vizino, Ezra: 159
Vizinus, Joseph: 3, 95, 98, 156–161

W (siglum defined): 53
Weekday: 25, 58–59, 67, 77, 81, 153–154, 156
William of Saint-Cloud. *See* Saint-Cloud, William of
Wright, R. R.: 87–89, 152, 173n, 181

Yaḥyā ibn Abī Manṣūr: 32, 131
Yuḥasin. *See Genealogies, Book of*

Zaragoza: 170
Zarzoso, M.: 170, 175
Zinner, E.: 4n, 90, 95, 181
Zúñiga, Juan de: 8, 47